D0164179

TOURISM IN NATIONAL PARKS AND PROTECTED AREAS

Planning and Management

Tourism in National Parks and Protected Areas

Planning and Management

Paul F.J. Eagles

*Department of Recreation and Leisure Studies,
University of Waterloo,
Canada*

and

Stephen F. McCool

*School of Forestry,
The University of Montana,
USA*

with contributions by

Elizabeth A. Halpenny

and

R. Neil Moisey

CABI *Publishing*

CABI *Publishing* is a division of CAB *International*

CABI Publishing
CAB International
Wallingford
Oxon OX10 8DE
UK

Tel: +44 (0)1491 832111
Fax: +44 (0)1491 833508
Email: cabi@cabi.org
Web site: www.cabi-publishing.org

CABI Publishing
10 E 40th Street
Suite 3203
New York, NY 10016
USA

Tel: +1 212 481 7018
Fax: +1 212 686 7993
Email: cabi-nao@cabi.org

A catalogue record for this book is available from the British Library, London, UK.

Library of Congress Cataloging-in-Publication Data
Eagles, Paul F. J., 1949–
Tourism in national parks and protected areas : planning and management / by Paul F.J. Eagles and Stephen F. McCool
p. cm.
Includes bibliographical references (p.).
ISBN 0-85199-589-6 (alk. paper)
1. National parks and reserves--Public use. 2. Protected areas--Public use. 3. National parks and reserves--Management. 4. Protected areas--Management. 5. Tourism--Management. I. McCool, Stephen F. II. Title.
SB486.P83 E24 2002
333.78′3--dc21 2002000681

ISBN 0 85199 589 6

Typeset by AMA DataSet Ltd
Printed and bound in the UK by Biddles Ltd, Guildford and King's Lynn

CONTENTS

CONTRIBUTORS

Paul F.J. Eagles, *Professor, Department of Recreation and Leisure Studies, University of Waterloo, Waterloo, ON N2L 3G1, Canada*

Elizabeth A. Halpenny, *Nature Tourism Solutions, R.R. #2, Almonte, ON K0A 1A0, Canada*

Stephen F. McCool, *Professor, School of Forestry, Montana Forest and Conservation Experiment Station, The University of Montana, Missoula, MT 59812-1063, USA*

R. Neil Moisey, *Assistant Professor, School of Forestry, Montana Forest and Conservation Experiment Station, The University of Montana, Missoula, MT 59812-1063, USA*

ABOUT THE AUTHORS

Paul F.J. Eagles is a Professor in the Department of Recreation and Leisure Studies at the University of Waterloo in Ontario, Canada. He is also a faculty member in the School of Planning and the Department of Biology. Professionally, Dr Eagles is a biologist and a planner who specializes in environmental planning. He has been involved in various aspects of park management for over 30 years, as a government employee, planning consultant, researcher and scholar with more than 270 publications. Since 1995, Dr Eagles has been Chair of the Task Force on Tourism and Protected Areas for the World Commission on Protected Areas. This Commission is part of the World Conservation Union (IUCN).

Paul F.J. Eagles

Stephen F. McCool is Professor of Wildland Recreation Management at the School of Forestry, The University of Montana in Missoula, Montana, USA. Dr McCool was initially trained as a forester, and then went on to receive MS and PhD degrees in outdoor recreation management. He has been professionally involved in management and planning of protected areas for over 30 years and has authored more than 200 publications. His work emphasizes sustainability, public participation and natural resource planning processes, particularly the Limits of Acceptable Change. He is a member of the World Commission on Protected Areas and serves on its Task Force on Tourism and Protected Areas.

Stephen F. McCool

Elizabeth A. Halpenny authored the chapter on tourism in marine protected areas. Elizabeth is a PhD student in the Department of Recreation and Leisure Studies at the University of Waterloo. Her work and studies are focused on tourism in protected areas, marine tourism and developing sustainable destinations.

R. Neil Moisey authored the chapter on the economics of tourism. He is an Assistant Professor of Nature-based Tourism in the School of Forestry, The University of Montana, in Missoula, Montana, USA.

PREFACE

Park management involves three interrelated aspects: (i) natural and cultural resources; (ii) visitors and the tourism industry; and (iii) the administering organization (Fig. P.1). The stewardship of the natural and cultural values which the park was established to protect receives fundamental attention and concern. These values produce a variety of benefits that attract visitors, both from the local area and from farther away. Planning and management of park visitors and tourism is a central concern as well. In order to manage the resources and the visitors there is an administrative organization – a corporate body and staff. These people and their organization must manage finances, human resources, legal aspects and political concerns that arise while stewardship is being practised and tourism managed. These three areas must all be considered simultaneously by any planning and management organization.

This book is primarily concerned with management of visitors and tourism. In order to properly implement this management, financial, staff, legal and political concerns are important, and are frequently discussed in the book. We place less emphasis on natural and cultural resource management, but these issues are certainly not ignored. The authors feel

Fig. P.1. Three aspects of park management.

that the planning and management of park visitation and tourism receives too little scholarly emphasis, thus the incentive for this book.

The central goal of this book is to describe the state-of-the-art of tourism planning and management in national parks and protected areas. A secondary objective is to provide guidelines for best practices in tourism operation. Other objectives are to:

1. Outline approaches for the planning and development of tourism infrastructure and services in national parks and protected areas.
2. Discuss the role of visitor management, including techniques that control and limit use so as to maximize visitor use while minimizing the negative environmental impact of that use.
3. Outline approaches for the enhancement of the quality of the tourism experience.
4. Describe case studies and guidelines for tourism that effectively contribute to the conservation of biological diversity and cultural integrity in these protected areas.
5. Describe case studies and guidelines on tourism in relation to the local communities resident within or near national parks and protected areas.
6. Provide guidelines on the measurement of tourism activity.
7. Provide a global focus to the subject matter.

The book takes a global perspective, with examples, case studies and photographs from many countries. Material is included from Argentina, Australia, Belize, Canada, Costa Rica, France, Ghana, Indonesia, New Zealand, Kenya, Mexico, Nepal, New Zealand, St Lucia, South Africa, Tanzania, the USA and the UK. The concentration is on countries that use English as their primary language and for publishing. The scholarship in park management is most heavily developed in the USA and Canada, and there is reliance on the literature from this region. However, the rapidly developing literature from other areas, most specifically Australia and the UK, is also strongly utilized.

The authors hope that the publication of this book will stimulate increased emphasis on and appreciation for the important area of park tourism. The world is approaching the end of the first century-and-a-half of rapid park expansion. The long period of management is beginning. As the emphasis shifts from establishment to management, it is critical that the planning and management of the visitation and tourism of national parks and protected areas be of the highest standard. As management approaches are utilized, they must be evaluated and the most successful retained and disseminated. Hopefully, this book will assist in the diffusion of information about the effective approaches already developed and in place. It may also stimulate further innovation in this important area.

CHAPTER 1
The Ecological and Cultural Goals of National Parks and Protected Areas

The Meanings of Parks

All parks are created by society for a purpose, which has varied across time and geography. Each park emerged within a particular societal ethos and organization. Over time, as the institution matured, different ideas of value came forward and some of these resulted in lasting landscape and management change. Others were more ephemeral, such as changing recreational fancy. Older parks, for example those of Central London, show over 500 years of use and ideas. It is very important for those who look with today's eyes and prejudices to understand the background of the landscape and cultures now observable.

Fig. 1.1. The protection of significant ecological values, economic development and provision of employment for local people are important themes of both the Kruger National Park and the adjacent private game reserves, such as the Sabi Sabi Private Game Reserve. Lion in Sabi Sabi Private Reserve, South Africa. (Photographed by Paul F.J. Eagles.)

One of the earliest definitions of the word 'park' is found in the *Oxford English Dictionary*:

> An enclosed piece of ground, of considerable extent, usually within or adjoining a city or town, ornamentally laid out and devoted to public recreation; a 'public park', as the various 'parks' in and around London, and other cities and towns. Also, an enclosed piece of ground, of considerable extent, where animals are exhibited to the public (either as the primary function of that 'park' or as a secondary attraction).

This ancient idea of an urban park is one of the oldest approaches, and one that has spread across the world with the movement of English peoples and culture. Over time this urban park idea was modified, expanded and transferred in many ways. In this section of the book the authors explore some of the central ideas underlying parks and their uses in society.

People visit parks with goals in mind. These goals are highly personal, but in mass also represent societal goals. It is important that park planners and managers understand the intentions of visitors.

To provide an initial idea of the many meanings of parks, we present a series of vignettes of the park experience. Each contains an illustrative account from the literature.

The theme of wilderness

The use of wilderness for personal reflection and redemption is a common theme, especially in the USA. This is an ancient biblical theme developed into a landscape and leisure phenomenon by the liberal Protestant Christian tradition in that country. In this theme, wilderness is a place away from normal life. It is a place to be alone, or with a small group. It is a place where nature is paramount, not people. There is danger in such a place, and each person must face this danger with a minimum of technology. It is place of reflection, a place that prepares a person for the challenges of normal life outside the wilderness.

The Bible is replete with references to wilderness. The word occurs 327 times in 42 different books. An example of the use of the concept of wilderness in the Christian Bible comes from Luke 4:

> 1. And Jesus being full of the Holy Ghost returned from Jordan, and was led by the Spirit into the wilderness,
> 2. Being forty days tempted of the devil. And in those days he did eat nothing: and when they were ended, he afterward hungered.
> Jesus was tempted by the devil to assume power over all things.
> 14. And Jesus returned in the power of the Spirit into Galilee: and there went out a fame of him through all the region round about.
> 15. And he taught in their synagogues, being glorified by all.
>
> (Luke 4: 1, 2, 14, 15)

This quotation shows the basic elements of the concept of wilderness. Jesus goes into the desert wilderness alone. He faces temptation and lack of food. His personal wilderness trials help him to prepare for the difficult challenges to come. He returns from the wilderness trials prepared to accept and surmount those challenges.

This theme has been adapted in North America to wilderness recreation, where individuals lead a spiritual quest into the wilderness, travelling alone or with a few companions. They take only a few supplies, eschew mechanized transport and accept nature on its own merits. They stay for long periods of time and accept the tests of nature on nature's own terms. They return from the wilderness psychologically strengthened to accept life's challenges.

The word wilderness comes from the ancient German phrase 'will doer ness', meaning a place of self-willed animals. 'Will' means self-willed, creatures not subject to the domination of people. 'Doer' means wild animal, and has come into English as deer, one type of wild animal. 'Ness' simply means place. Therefore, a wilderness is a place where all of nature exists of its own accord, where humans are secondary and must not impose their will. The word is well understood in English, German, Dutch and Scandinavian languages. The use of the word wilderness in the earliest translations of the Bible had profound cultural impact on those societies. Interestingly, no similar word exists in French or Latin or in the

Romance languages derived from Latin and, as a result, there is much less understanding of the concept of a wild area free from human intervention.

In most usage, the Biblical connections with wilderness recreation are not remembered. However, the underlying concepts are alive and powerfully used by the intelligentsia of the USA and, to a lesser degree by those of other English-speaking countries. The USA was the first country to place the concept into legislation, with the passage of the Wilderness Act in 1964. This Act states that wilderness is 'an area where the earth and its community of life are untrammeled by man, where man himself is a visitor who does not remain' and 'has outstanding opportunities for solitude or a primitive and unconfined type of recreation' (USFS, 2001). In this Act the concepts of wild lands untrammelled by people but used for solitude and reflection are ensconced in law. The US Wilderness Areas system now contains 633 areas covering 42.9 Mha (106 million acres) (G. Marangelo, 2001, personal communication).

Community social function

Parks are areas for community events, social functions and athletic competitions. Since Roman times, European city squares and plazas have fulfilled vital community social functions. In the southern European tradition, cities have formal squares with the church, government offices, the police and sellers of wares on the four edges of the square. This tradition has been transplanted into most cities of Latin America. The northern European tradition has similar spaces with similar purposes, but they are typically greener and less formal.

As an example of the use of city squares for community social functions we follow this theme from James Michener's book, *Texas*. In the year 1716 . . .

> At age 26 Simon was undergoing an experience in this northern Mexican town which disturbed him and at the same time delighted him. In the past his occupation had kept him on the move and a lack of money had prevented him from paying court to the young women in those towns where he worked, but in Zacatecas he had steady employment, so 6 nights a week when work was done he found himself in the spacious public square before the cathedral, watching as the young unmarrieds of good family walked about from 7 to 9.
>
> They did not walk aimlessly. The men strolled unhurriedly in a counterclockwise direction, keeping toward the outside of the tree-lined square, and as they went they looked always toward the center of the square, where inside the large circle they had formed, walked the young women of the town in a clockwise mode. About every ten minutes a young man would meet head-on, almost eye to eye, a particular young woman, twice in each circuit of the plaza, and in this practical, time-honoured Spanish manner the unmarrieds conducted their courtships. Over a period of 3 weeks, any young

> man could pass his preferred young woman more than a 100 times, during which he could notice with the precision of a scholar the degree to which her smiles had softened.
>
> (Michener, 1985, p. 74)

This story shows the role of the central square of Mexican cities in one community social function, that of courtship. Here young men and women could meet in a socially acceptable place and circumstance. This place was created by society to provide a safe and orderly setting for this and many more important social needs.

Fig. 1.2. Hotsprings in central England were developed into baths by the Romans and used continuously from AD 65 to 410. For millennia, societies have created special places that fulfil vital community social functions. Natural hot waters for bathing and social activity were important to the Romans, and this attraction was a major element in the creation of three of the earliest national parks in North America, the Arkansas Hot Springs, Yellowstone and Banff. Roman Baths World Heritage Site in Bath, England. (Photographed by Paul F.J. Eagles.)

Parks fulfil many social functions, such as courtship, family bonding, community cohesion, athletic competitions and the meeting of people.

Hunting preserve

Since the Middle Ages in Europe, male members of royalty have set up hunting reserves which were managed for the personal recreation of the upper classes of society. The *Oxford English Dictionary* gives one definition of a park that fits this approach: 'An enclosed tract of land held by royal grant or prescription for keeping beasts of the chase. (Distinguished from a *forest* or *chase* by being enclosed, and from a *forest* also by having no special laws or officers.)' This use of the word extends as far back as 1260. Typically, commoners were excluded from Royal Parks, except as gamekeepers and servants. Over time some commoners obtained rights of access for use, such as the collection of grass for thatched roofs or the grazing of sheep. When these common people emigrated elsewhere in the world, they took with them the idea of hunting reserves. They often set them up in their new countries, but in modified forms. In the New World the hunting reserves were created but an important innovation allowed access for all people.

The initial reserve creation for Banff National Park was undertaken by the national government of Canada in 1885. The first management report on the new reserve was written in 1886. In that report the author, W.F. Whitcher, saw the potential of Banff for hunting and profit.

> There are recreative and attractive features about the prevalence of edible game in every new country that become in fact of the highest and most profitable utility, and which the progress of settlement and growth of trade serve greatly to enhance.
>
> (quoted in Foster, 1978, p. 29)

This report's ideas on hunting were not accepted by the national government in the emerging concept of national parks in Canada. However, later, hunting reserves, called Wildlife Areas in Canada and Wildlife Refuges in the USA, were established. The National Wildlife reserve systems are now large and heavily used for hunting. The US system includes 530 refuges, covering 37.7 Mha (93 million acres) (USFS, 2001). The Canadian System is much smaller with 48 areas covering 0.5 Mha (1.2 million acres) (Burns and Warren, 2000).

Business and profit

Since the beginning of parks, business and profit has been an attractive element to some sectors of society. When people visit an area there is a potential for the sale of goods and services. When large numbers of people visit from a distant area, the concept of tourism develops. In the early 1880s Sir Sandford Fleming, a railway engineer, proposed a system of parks across Canada for the purpose of attracting tourists who would travel on the new Trans-Canada railway. Sir Sanford Fleming . . .

> . . . made the first proposal for a national park in Canada, in fact two parks, one at Lake Superior and one in the mountains. His motive was far from altruistic. He saw the Rockies as another Switzerland, 'a source of general profit,' especially for the CPR, which could carry the tourist traffic. In a few short paragraphs published in his book *England and Canada* (1884), Fleming predicted the kind of wealthy patrons such a park would attract, and mentioned the improvements, the bridle paths and 'retreats' that would be needed for them. It was a small but accurate blueprint for the first park, whose birth was drawing very near.
>
> (Marty, 1984, p. 32)

The support from powerful business interests was a critical element in the creation of many parks. In both the USA and Canada, the powerful railway lobbies were strong supporters of the creation of many of the first national parks. The support of tourism interests is still vital for the continued political survival of many parks.

Physical and emotional health

Parks are special places for the restoration of the physical and emotional health of visitors. They help to renew a person's health and relieve the stress of urban living. Access to open air, sunshine and nature is seen as healthful. For example, since Roman times, there has been a fascination with hot mineral waters for bathing. Three of the earliest parks in North America were set aside around natural hot springs: Arkansas Hot Springs in 1832, Yellowstone National Park in 1872 and Banff National Park in 1885. For centuries people travelled long distances to immerse themselves in hot springs, largely for health reasons.

As a result of vigorous lobbying by public-spirited individuals interested in the health and welfare of the average person, in 1851 Toronto was the first city in Canada to establish a parks and recreation management agency. After 8 years of operation, Toronto's Committee on Public Walks and Gardens, the descriptive name of Canada's first park management agency, was prepared to take political action to forward the idea of urban parks. Interestingly, the healthful attributes of urban parks were prominent in a speech made by the chairman of the committee to a Toronto City Council meeting in 1859. In that speech he outlined to the city council his idea for urban parks:

> In the first place, they furnish to the wealthy places of agreeable resort, either for driving or walking, and free from exposure to the heat and dust of

an ordinary road . . . thus enabling them to enjoy the inestimable blessing of the free open air of the Country – so conducive to the promotion of health and morality . . .

In the second place, to the mechanic and working classes, Public Grounds are of incalculable advantage. How much better it is for the families of such to have these places of recreation and healthful exercise, than to have them exposed on the crowded streets of the city?

What more pleasing sight to the philanthropic mind than to witness the wholesome rivalry of the mothers of families, on the Sabbath or the weekday 'Summer eve' as to whose children shall appear the cleanest and neatest clad. How are such, and kindred inspirations, calculated to elevate and refine the mind, and improve the condition of all.

(McFarland, 1982)

Fig. 1.3. On discovery, the hotsprings in the Rocky Mountains were the initial site reserved for protection by the Canadian government in 1885. The area was quickly expanded and ultimately became Banff National Park, Canada's first national park. Cave and Basin Hotsprings in Banff National Park, Canada. (Photographed by Paul F.J. Eagles.)

As can be seen from this speech, the committee assumed the responsibility for providing 'public grounds' for all classes of society, and especially for the working class who badly needed access to 'places of agreeable resort', provided free of charge by society. This public-spirited and socialistic approach to parks, public use subsidized by community taxes, became a fundamental aspect of park management in most developed countries.

Contact with wild nature is seen as restorative for mental health, as shown in the poem, *The Peace of Wild Things*, by Wendell Berry.

> When despair for the world grows in me
> and I wake in the night at the least sound
> in fear of what my life and my children's lives may be,
> I go and lie down where the wood drake
> rests in his beauty on the water, and the great heron feeds.
> I come into the peace of wild things
> who do not tax their lives with forethought
> of grief. I come into the presence of still water.
> And I feel above me the day-blind stars
> waiting with their light. For a time
> I rest in the grace of the world, and am free.
>
> (Bly, 1980, p. 179)

The idea that parks provide physical and mental health benefits is still a strong theme in park management.

Ecological preservation

Much early park preservation was not system-oriented in scope. It concerned the preservation of unique geographic features, such as ravines or waterfalls, or wildlife. The earliest expressions of landscape preservation concerned the protection of the productive capabilities of ecosystems. For example, the creation of Algonquin National Park by the Province of Ontario in Canada in 1893 was to preserve the forests for sustainable logging and to preserve the headwaters of the many rivers that flow out of the Algonquin highlands. In the late 1940s, some innovative literature began to represent an ecosystem view of park creation and management. It was not until the 1960s, with the emergence of the science of ecology, that concepts such as endangered species or ecological planning emerged fully fledged.

In the UK in 1947, a national government white paper proposing a set of National Nature Reserves stated that they shall be:

> to preserve and maintain as part of the nation's natural heritage places which can be regarded as reservoirs for the main types of community and kinds of plants and animals represented in this country, both common and

> rare, typical and unusual, as well as places which contain physical features of special or outstanding interest.
>
> (quoted in Mabey, 1980, p. 218)

This call for the representation of 'the main types of community' is an early statement of the need for a reserve system across a country to protect representative examples of natural areas. This report led to the development of sites of special scientific interest and the National Nature Reserve System in the UK. Later, ecological representation became a major force in the creation of systems plans for large geographical areas such as nations, states and provinces. Today, ecological system planning is a fundamental component of many park systems. Under this approach, parks are created with an eye to their role within the ecosystem. In the 1990s, as the ecological disadvantages of isolated habitats became obvious, the linkage of parks by corridors of natural habitat became a major focus of activity.

One of the disadvantages of the ecological preservation theme is the frequent lack of recognition that parks have been and still are important for other ideas. Much revisionist history has been written where modern ecologists try to shoehorn older ideas and parks into a more recent concept of reality.

Recreation

Recreation is an important use of parks and protected areas. The concept has several meanings.

The *Oxford English Dictionary* provides a definition of park that shows the connection between a pastoral landscape, land ownership and recreation common in England and widely transported elsewhere: 'Hence extended to a large ornamental piece of ground, usually comprising woodland and pasture, attached to or surrounding a country house or mansion, and used for recreation, and often for keeping deer, cattle, or sheep.' This pastoral landscape of pasture lands, mixed woodlands and large country homes is the model for city recreation commonly used throughout much of Europe and North America. The recreation done there is that of the civilized, the cultured and the privileged. Hence the extensive grassy swards so common in urban parks.

Recreation is the act of creating over again, of renewing, of replenishing. What are we looking for when we recreate in the countryside? Richard Mabey proposes that:

1. We prefer open country to fenced;
2. Access to exclusion;
3. Variety to monotony;
4. Stability to change; and

5. Living things to inanimate structures.

(Mabey, 1980, p. 162)

Therefore, in Mabey's approach, recreation in rural landscapes is stimulative of the mind and inherently conservative.

A traditional use of the word recreation means a sporting event, an event with considerable physical activity. Accordingly, parks are heavily used for sport, both organized in the form of competitive games and unorganized in the form of family fun. Occasionally, the demand from sport requires that parks are heavily modified with specialized facilities. Running tracks, grassed pitches and horse stables are constructed. Many of the oldest of the urban parks have a cornucopia of such sport and recreation facilities.

Outdoor recreation in parks has a long and deep tradition stretching back 500 years or more in Europe. Such recreation has occasionally been at the forefront of social change. In 1906, Fancy Case started the world's first outdoor camp for girls in Algonquin National Park in Ontario. She had firm ideas on the goals of such a park, as outlined in her annual advertising booklet for Northway Camp.

> Restful, homelike conditions prevail, the tents being placed far apart and composed of four in a family – a counsellor and three girls.
>
> Schedule and competition are made light of. Time 'to think of and revalue the durable satisfactions of life' seems more important.
>
> Self-help and democratic ways are favourite ways, all having a share in the management of and contributing to the carrying out of the camp ideals. Camp properties belong to all equally.
>
> Individual, not mass life, and a full but free environment for the purpose of satisfying our different tastes and abilities, are goals. Camp is naturally an ideal place for developing social responsibilities and consciousness.
>
> Fun, play and happiness are frankly aims of our summer vacation camp life as well as love for work and service.
>
> A belief in work. The hour or two of real work, mornings before swimming, is connected with building small cabins, clearing out dead trees, cleaning up the forest, blazing new trails, fixing up old trails, painting and mending boats and other necessary pieces of work.
>
> (quoted in Raffin, 1999, p. 118)

This heady bit of feminine liberation philosophy in the guise of outdoor recreation was taking place in a lakeside camp deep in a Canadian park, almost two decades before Canadian women were legally recognized as 'persons' and given the right to vote.

Meaning of life

Wild nature is often used to reveal life's meaning. The concepts underlying nature can be used to guide human lives and human systems.

> While nature is not a uniquely suitable setting, it seems to have a peculiar power to stimulate us to reflectiveness by its awesomeness and grandeur, its complexity, the unfamiliarity of untrammeled ecosystems to urban residents, and the absence of distractions. This special additional claim for nature as a setting is that it not only promotes self-understanding, but also an understanding of the world in which we live . . . Nature is also a successful model of many things that human communities seek: continuity, stability and sustenance, adaptation, sustained productivity, diversity and evolutionary change.
>
> (Sax, 1980, p. 47)

Parks and the experiences coming from parks provide an important meaning of life for many people. This meaning may be the most important souvenir retained by most visitors.

Protecting native people and their lands

In 1832, George Catlin, an artist, travelled up the Missouri River into the central plains of North America. This was the home of plains Indian people. On his return to the USA in 1833 he wrote a letter to the *Daily Commercial Advertiser* in New York City and stated that these regions:

> might in future be seen (by some great protecting policy of government) preserved in their pristine beauty and wildness, in a magnificent park, where the world could see for ages to come, the native Indian in his classic attire, galloping his wild horse amid the fleeting herds of elks and buffaloes. What a beautiful and thrilling specimen for America to preserve and hold up to the view of her refined citizens and the world, in future ages. A nation's Park containing man and beast, in all the wild and freshness of their nature's beauty.
>
> (Huth, 1957, p. 135)

Interestingly, this first call for a national park was for the protection of an aboriginal people and the lands that sustained them. This was an inherently ecological idea; people lived on the land and needed the preservation of those resources to survive. The US government did not accept this idea. Instead, it sent in the cavalry to destroy the Native Americans and to steal their lands.

The idea of a great protecting policy by the national government for lands in a nation's park later took root with the establishment of the Mariposa Grove of giant sequoias as Yosemite State Park in 1864 and Yellowstone National Park in 1872. Significantly, in the creation of these parks no native peoples were allowed to stay in the parks or were allowed to keep a right of ownership.

However, times have changed with regard to aboriginal rights and parkland. Some protected areas contain human populations, many that are aboriginal. In Australia, some of the national parks are owned by the aboriginal peoples and are leased to the national government for management. Uluru-Kata Tjuta National Park and Kakadu National Park both have this administrative situation. The management plan for Uluru-Kata Tjuta is based on the principle of aboriginal ownership and involvement:

> Many places in the Park are of enormous spiritual and cultural importance to the traditional owners. The Park also contains features such as Uluru and Kata Tjuta, which have become major symbols of Australia. Acknowledgment of Uluru-Kata Tjuta as a cultural landscape is fundamental to the success of the joint management arrangement. The Park is managed in such a way that the rights, interests, skills and knowledge of the traditional owners are respected and integrated in all of the Park's management programmes.
>
> (Parks Australia, 2001)

In Canada's north, the Inuit peoples often retain access and use rights in recently established national parks. In many countries, aboriginal reserves, that is native peoples' homelands, are treated as special forms of protected areas. For example, Costa Rica maintains that 28% of its lands are in protected status, in three types of protected areas: national parks, wildlife reserves and aboriginal reserves.

Historical and cultural preservation

Many countries created special designations and management structures for important historical and cultural sites. The preservation and

Fig. 1.4. Many countries have special designations for important historic and cultural sites. Throughout Central America, Mayan cities are protected and commemorated with park designation. El Castillo, Mayan City of Chichen Itza, Mexico. (Photographed by Paul F.J. Eagles.)

interpretation of these places significant in history has high national priority. These places often preserve important historical sites, landscapes, buildings and artefacts. Canada and the USA have integrated their national historic preservation programmes into their national park management agencies. In both these countries, there are hundreds of special sites protected and managed for their historical significance. In the UK, a private body, The National Trust, carries out this responsibility.

A unique example of a historic site is Vimy Ridge in France. This is the site of an important Canadian victory over the Germans in the First World War. A reading from Pierre Burton's book on the battle gives a feeling of what occurred there.

> In all of history no human ears had ever been assaulted by the intensity of sound produced by the artillery barrage that launched the Battle of Vimy Ridge on April 9, 1917.
>
> In the years that followed, the survivors would struggle to describe that shattering moment when 983 artillery pieces and 150 machine guns barked in unison to launch the first British victory in thirty-two months of frustrating warfare. All agreed that for anyone not present that dawn at Vimy, it was not possible to comprehend the intensity of the experience. The shells and bullets hurtling above the trenches formed a canopy of red-hot steel just above the heads of the advancing troops – a canopy so dense that any Allied airplane flying too low exploded like a clay pigeon. At least four machines were destroyed that morning by their own guns.
>
> The wall of sound, like ten thousand thunders, drowned out men's voices and smothered the skirl of the pipes – the Highland regiments' wistful homage to a more romantic era.
>
> Tons of red-hot metal hurtling through the skies caused an artificial wind to spring up, intensifying the growing sleet storm slanting into the faces of the enemy.
>
> The barrage began exactly at 5:30 A.M. Technically, it was dawn, but the first streaks of light in the east were obliterated by the driving storm. Shivering in the cold, tense with expectation, their guts briefly warmed by a stiff tot of army rum, the men in the assault waves could scarcely see the great whale back of Vimy Ridge, only a few hundred yards away. It angled off into the gloom – its hump as high as a fifty-storey building, a miniature Gibraltar, honeycombed with German tunnels and dugouts, a labyrinth of steel and concrete fortifications, bristling with guns of every caliber.
>
> The Germans had held and strengthened this fortress for more than two years and believed it to be impregnable. The French had hurled as many as twenty divisions against it and failed to take it. In three massive attacks between 1914 and 1916 they had squandered one hundred and fifty thousand dead or mangled. The British who followed the French, had no better success. Now it was the Canadians' turn . . .
>
> The Canadian Corps (which included one British brigade) faced an incredible challenge. In one day, in fact in one morning, these civilian volunteers from a small country with no military tradition were expected to do what the British *and* French had failed to do in two years. The timetable called for most of them to be on the crest of the ridge by noon. And they

> were expected to achieve that victory with fifty thousand fewer men than the French had *lost* in their own frustrated assaults.
>
> Few thought they could succeed. The Germans didn't believe that any force could dislodge them. A few days before the battle, one confident Bavarian put up a sign reading: 'Any body can take Vimy Ridge but all the Canadians in Canada can't hold it.' A German officer taken in a raid before the battle told his captor: 'You might get to the top of Vimy Ridge but I'll tell you this, you will be able take all the Canadians back in a rowboat that get there.'
>
> Well the Canadians did take Vimy Ridge that day in June. This was the first major victory for the Allies in World War I, and it was done by Canadian volunteers.
>
> (Burton, 1986, pp. 14–21)

Vimy Ridge is now a Canadian national historic site in France, managed by Veteran Affairs Canada with the assistance of Parks Canada. This First World War battlefield site includes landscape features such as trench systems, tunnels, shell holes and mine craters. It has permanent staff, interpretive programmes and regularly scheduled commemorative events. This is a rare national historic site that is designated by one country, Canada, within another country, France.

Overall, Canada has designated 849 national historic sites, 557 persons and 324 other aspects of Canadian history. The most common form of commemoration is by a plaque or a simple marker. Places designated as national historic sites are occasionally acquired by the federal government for protection and interpretation. Of the 849 national historic sites, Parks Canada administers 145 and contributes money to an additional 71 managed by other governments or organizations (Parks Canada, 2000).

Summary of the Meanings of Parks

Parks and protected areas represent a rich and complicated suite of ideas. Park managers must be fully aware of the history of the meanings contained in any one site, as well as the changes in emphasis over time. The oldest parks have been swept by changing concepts many times and, as a result, contain a complex assemblage of landscapes, artefacts, structures and landforms. A walk through Hyde Park in London will reveal to the keen observer the many societal ideas that have flowed into this park over the 500 years of use.

It is critical that all those involved in parks – the managers, the visitors, the lobbyists – recognize and respect the range of ideas involved. Conflict, which is typically caused by goal interference, stems from different ideas of what is desirable and acceptable in parks. Many societal values occur in parks. The processes used to represent these values are key to parks' societal relevance.

The Assignment of Value

The determination of the value of a park and its components is critical. Any activity, human use or impact can be considered to be negative, neutral or positive depending on several factors, such as the point of view of the observer, the time of year or the costs and benefits derived. It is therefore important to recognize that value assessment is fundamentally a political, not a scientific, process.

The determination of value is a major part of planning and managing parks. There are many key groups that play roles in the determination of impact value. Who assigns value and how the value is assigned are central issues. Also critically important is the method used to assess this value.

A typical approach is to *let the park staff assign value*. Most park agencies have highly trained staff with years of education and experience. They often know a lot about the environment and about human culture. They are familiar with the existing agency policies. They typically have specialized and highly technical knowledge. They are emotionally involved with the park. They often feel very strongly about the significance of various park resources. They often feel they are ideally suited for the task. It is very common for park staff to have the central role of determining the values assigned to park resources.

Sometimes the decision is made to *let independent experts assign value*. Most societies contain much expertise in many fields. All aspects of natural, social, economic and cultural resource issues are known to highly trained people in universities, in schools, in government and in industry. These people have valuable information and insight that can be brought to bear on any issue. Sometimes, they are given the role of value determination because of their strong knowledge.

Another approach is to *let the politicians assign value*. In democratic systems, all park managers have an elected political master. These people were elected to represent a group of people in a ward or a riding. They are popular in their community. They were elected to carry out certain policies and are quite familiar with the views of their constituents. They are often in the best position to do the job.

Occasionally, it is best to *let the local community assign value*. The people immediately around the park are directly affected by all park policies. Often other resource uses, such as those that are extractive, are forgone with the establishment of the park. The park affects the local economy, as park visitors travel through, visit and have an impact on the local community. The local people may know the park area well. They have probably lived there for a long time and have seen nature in its many manifestations. They usually demand a say in any policy and are very willing to assign the value to all resources and developments.

Rarely, a decision is made to *let the park visitors assign value*. The visitors are keenly interested in the park. They have taken their valuable leisure time to come to experience the park. They pay for the privilege of

Fig. 1.5. The trail to the volcano, in Volcan Poas National Park, goes through the visitor centre, enticing all visitors to learn about this park and the other parks in the national system. Visitor centres help to inform local people and others of the values represented by parks. Visitor centre at Volcan Poas National Park, Costa Rica. (Photographed by Paul F.J. Eagles.)

visiting. In older parks they may have visited the parks over very long periods and have developed unique perspectives. A visitor is often very appreciative of the park and its unique features. Visitors are very willing to give their opinions and often demand to be heard. Some might argue that the park has been established for the use of the visitors. They have a unique position from which to determine resource value.

The potential park visitors form an important group. There are many people inclined to visit the park who have not yet been able to do so. It is distinctly possible that the park policies are creating an impediment to their visit. Maybe the fees are too high. Maybe the activities that they most desire are not allowed. Maybe they are physically challenged and need special facilities. Maybe they want a more relaxed atmosphere and fewer crowds. They might want to encourage certain species of wildlife. Such people are often interested in providing their ideas on how the natural resources should be valued and managed. The *potential park visitors might wish to assign value.*

Senior governments run some parks. These parks cater to a wide geographical area. The argument can be made that park policies should reflect the view of the people across the entire jurisdiction of the government. For example, national park policies should reflect the policies of the entire country. Many parks have resources of worldwide significance and are, in essence, important to all people. Therefore, possibly *all people should have a say* in the assignment of value to park resources.

In practice, many constituencies influence park management. Natural and cultural resource decision making must be considered within such a context. A decision-making system must be developed that realistically and effectively provides an opportunity for all people in all constituencies to participate. No single group should be allowed to dominate.

It is worth mentioning that the soundness of all decisions is heavily influenced by the values and knowledge of the involved public. It is therefore critically important that the ecological and cultural roles of parks are communicated to and understood by the public.

In reality, every major decision in parks is ultimately subject to a formal political process. How many people support a particular decision is the telling point for democratic governments. All park managers must be aware of this fact.

The assessment of impact is dependent on the determination of value within an overtly political process. The park visitors are one of the least influential groups in decision making. They visit for a short period, often live far away from the park and are poorly organized politically. It behoves park officials to develop procedures to ensure that park visitors and potential park visitors are given a voice in park decision making.

The IUCN Definitions of National Parks and Protected Areas

Every park jurisdiction has a terminology and a management structure for its parks. There is often mimicry of other systems and ideas, but local political conditions often lead to a unique twist. The United Nations requires that a list of the world's national parks and protected areas is tabulated and kept. The compilation of such a list requires standard definitions of such terms as national park, nature reserve and wildlife refuge. Accordingly, a standardized set of terminology and definitions was developed, known as the IUCN Category System for National Parks and Protected Areas (Table 1.1). This system is ecologically based. Category I parks have the highest level of ecological integrity, with the least level of human impact. As one goes down the categories from I to VI, the amount of human interference in ecology gets greater. Category V parks are protected landscapes, much like British National Parks. These are human- modified landscapes with regional planning control. Category VI sites are managed landscapes often used for recreation and resource extraction such as logging.

This classification system does not work well with cultural sites or historic parks because of the classification's underlying emphasis on ecological values. Therefore, historic parks are not included in the United Nations' inventory of national parks, even though they are called national historic parks in many countries and are managed by the national park agencies. This classification ignores urban parks as well, unless they are large and have high ecological integrity.

This classification does not explicitly recognize the role of visitors or tourism in parks, except in a backhanded, negative fashion. Those sites

Table 1.1. IUCN's Category System for National Parks and Protected Areas.

Category I	*Strict Nature Reserve/Wilderness Area*: Protected area managed mainly for science or wilderness protection
Category IA	*Strict Nature Reserve*: Protected area managed mainly for science
Category IB	*Wilderness Area*: Protected area managed mainly for wilderness protection
Category II	*National Park*: Protected area managed mainly for ecosystem protection and recreation
Category III	*Natural Monument*: Protected area managed mainly for conservation of specific natural features
Category IV	*Habitat/Species Management Area*: Protected area managed mainly for conservation through management intervention
Category V	*Protected Landscape/Seascape*: Protected area managed mainly for landscape/seascape conservation and recreation
Category VI	*Managed Resource Protected Area*: Protected area managed mainly for the sustainable use of natural ecosystems

with the least visitation are given the highest ranking. However, Category II parks, typically called national parks, often have very high visitation levels due to their exceptional natural resources and high public profile.

A good way of considering the role of visitation within the classification system is to think about the gradation of the types of activities allowed. From Category I to Category VI, human activities become more intrusive. Category I sites have people visiting in low numbers with a minimum of infrastructure and a minimum of technological interference. Category II and III sites allow tourism infrastructure, such as roads, visitor centres and campgrounds, in a small part of the park. Category IV sites allow consumptive recreation, such as hunting. Category V allows considerable levels of human intervention, such as farming, houses and extensive tourist facilities. So does Category VI, with the addition that VI allows all manner of extractive activities including mining, forestry, commercial fishing and a whole suite of mechanized recreation.

There are park classification systems in use that give more explicit recognition to historical and tourism aspects than does the IUCN classification. One of these is used by Ontario Provincial Parks in Canada.

The Ontario Provincial Parks Classification System

The Province of Ontario in Canada has an old and well-developed park system. In 2001 the system contained 275 parks covering 7,100,000 ha. In addition, 378 new parks covering 2,400,000 ha are promised and being established over time (Ontario Ministry of Natural Resources, 2001). These parks and protected areas are managed within a seven-class system (Ontario Ministry of Natural Resources, 2000). Each class has a unique focus and management regime (Table 1.2).

Table 1.2. Ontario Provincial Park Classes.

Class name	Definition
Wilderness	Wilderness parks are substantial areas where the forces of nature are permitted to function freely and where visitors travel by non-mechanized means and experience expansive solitude, challenge and personal integration (IUCN Category IB)
Recreation	Recreation parks are areas that support a wide variety of outdoor recreation opportunities for large numbers of people in attractive surroundings (no IUCN equivalent)
Natural environment	Natural environment parks incorporate outstanding recreational landscapes with representative natural features and historical resources to provide high quality recreational and educational experiences (IUCN Category II)
Historical	Historical parks are areas selected to represent distinctive historical resources of the province in open space settings and are protected for interpretive, educational and research purposes (no IUCN equivalent)
Nature reserve	Nature reserves are areas selected to represent the distinctive natural habitats and landforms of the province, and are protected for educational purposes and as gene pools for research to benefit present and future generations (IUCN Category IA)
Waterway	Waterway parks incorporate outstanding recreational water routes with representative natural features and historical resources to provide high quality recreational and educational experiences (no IUCN equivalent)
Conservation reserve	Conservation areas protect representative natural areas and special landscapes while allowing consumptive recreation, such as hunting (IUCN Category IV)

The Ontario classification system has some similarities to the IUCN classification. It has wilderness, nature reserve, natural environment and conservation reserve categories that are equivalent to IUCN Categories IB, IA, II and IV, respectively. Ontario has classes that are not found in the IUCN system. The historical class contains historical and cultural resources for preservation and interpretation purposes. Parks with a cultural and historical focus are common in the USA and in Canada, but have no place in the ecologically based IUCN classification. The Recreation Class contains landscapes that provide for significant levels of outdoor recreation activity. There is no IUCN equivalent. The Waterway Category is especially suited for the Ontario landscape with its abundant lakes and wild rivers. This system is specifically aimed at protected landscapes that contain outstanding water-based recreation opportunities, in which long-distance wilderness canoeing is a prominent activity. This recreation-based category also has no IUCN equivalent.

The Ontario classification system is featured because of its explicit recognition of the role of outdoor recreation, history and tourism in provincial parks. With this system there is no need to try to shoehorn activities into an inappropriate class, as so often occurs with the polyglot Category II, National Park, in the IUCN Category System. It would be worthwhile if outdoor recreation and tourism were more explicitly recognized in the global park classification system of the IUCN.

The Status of the World's Parks

The data collection for the United Nations' List of National Parks and Protected Areas allows for the development of an understanding of park creation over time. Figure 1.6 shows the growth of the global system of protected areas over a 100-year period. Both the number of parks and the area of these parks grew substantially over time. The growth curve increased in 1960, and this high level of growth continued until the present. By 1996 the world's network of 30,361 parks covered an area of 13,245,527 km^2, representing 8.84% of the total land area of the planet. This total land area spans 225 countries and dependent territories (Green and Paine, 1997).

There is no global inventory of park tourism. None has ever been compiled. Realizing the importance of such data, the World Conservation Monitoring Centre, in close cooperation with the World Commission on Protected Areas, made a first attempt at collecting global park-use data during the 2002 data collection for the next edition of the United Nations' List of National Parks and Protected Areas. The publication of these data is expected in 2003.

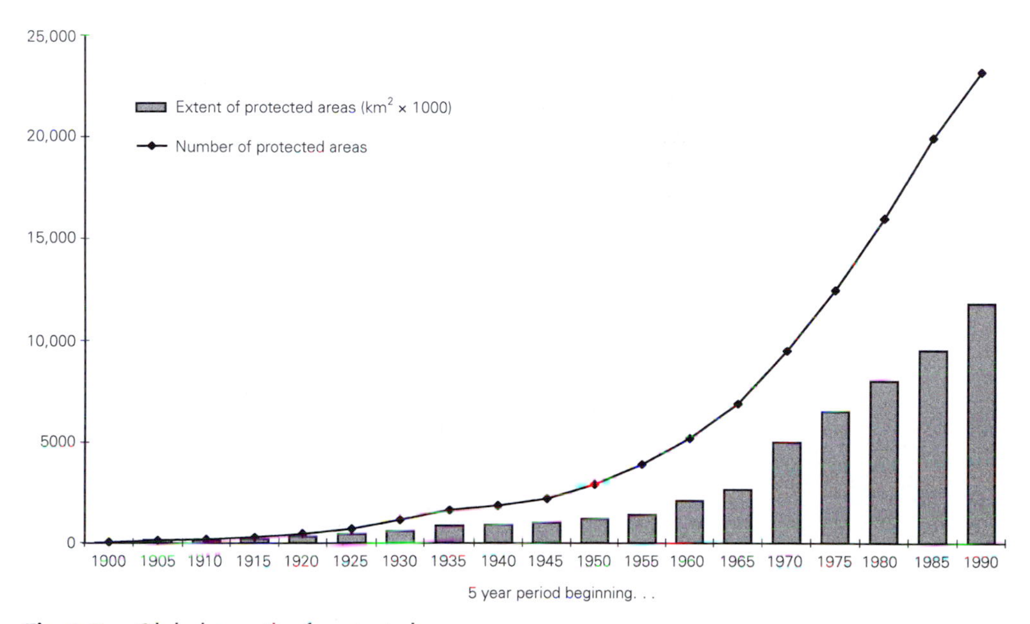

Fig. 1.6. Global growth of protected areas.

Biological and Ecological Conservation

Starting in the 1960s, the ecological ethic became increasingly important in the establishment of parks and in their management. This change occurred along with a general societal recognition of the importance of ecosystem conservation.

A general concept underlying ecological conservation is that of higher ecological integrity in the absence of human interference. The highest level of integrity is that which occurs when people are not present and do not interfere with natural ecological processes. Therefore, the benchmark from which all impacts are measured is that of no human presence and no human impact. This is a high standard to reach since virtually all ecosystems in the world have been impacted by humans for periods ranging from thousands to millions of years. All humans came from natural ecosystems and depended on natural ecosystems for all of life's necessities. However, most parks and protected areas in IUCN Categories I to IV involve landscapes where humans are not allowed to live permanently, and in Categories I to III are not allowed to have a material impact on the natural ecosystems to any significant degree. Therefore, the world's national parks and protected areas are extremely important in their benchmark roles. They provide virtually the only areas on the Earth's surface where natural ecosystems occur and can be studied, with minimum negative human impact.

The Interrelationships between Conservation and Tourism

There is a common sense concept that human impact on parks and protected areas is inherently negative. This flows from the observation that when humans enter a Category I, II or III park, they change the system that occurs in their absence. This naturally leads to the conclusion that all human activities in parks are interfering and damaging.

This concept is shallow. It does not recognize that it is human action that leads to the creation of a park, and it is ongoing human activity that establishes a management regime that protects the ecological and cultural values of a park. In the absence of the legal actions of creation and management, the landscapes would be used for some other activity.

The creation and management of a park is a political action. It happens when a government has sufficient public and private interest to undertake the legal and political action of park creation. Governments are always under pressure to make changes and to propose laws to their legislative body. Any one action only occurs when it has higher priority than other competing actions. Therefore, governments and legislatures only create parks and provide resources for their management when a sufficiently large and influential group of people want such an action.

Fig. 1.7. The active Poas volcano, in Volcan Poas National Park, attracts large numbers of visitors from the capital city of San Jose, Costa Rica. This provides park managers with an opportunity to introduce all visitors to the concept of national parks throughout the country. Governments and legislatures only create parks and provide resources for their management when a sufficiently large and influential group of people wants such an action. Weekend crowds at Volcan Poas National Park, Costa Rica. (Photographed by Paul F.J. Eagles.)

There are several fundamental cultural features that must be present before a critical mass for parks occurs. First, a societal attitude must be present that recognizes value in parks, typically ecological and cultural conservation as well as recreation demand. This attitude must be sufficiently strong to make citizens act. They must propose a park. They must lobby for the park. They must influence other citizens. They must influence political leaders. They must be prepared to pay with time and money. Second, this value must be as strong or stronger than other competing values. Government has only so much time and money. Other competing interests include health care, education, the military and societal infrastructure. The park values must be strong enough to compete successfully with these other interests in the halls of power. Society at large must be prepared to accept this institutional change of park creation and management. It must work around this activity and be prepared to make changes in other demands.

Park visitation is critical to the creation of societal culture conducive to parks. People must visit parks, must appreciate the experiences gained and must have a memory of appreciation that leads to long-term attitude reinforcement. They must develop a sufficiently strong attitude that causes political action towards parks. Only when sufficiently large numbers of people gain such attitudes and take such actions do governments see the need to move.

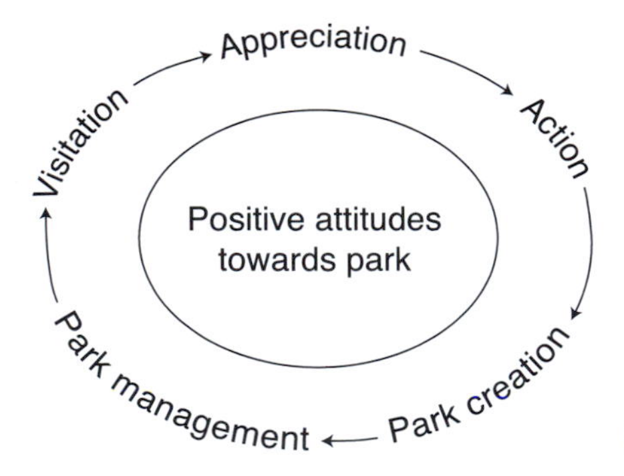

Fig. 1.8. Tourism and conservation cycle.

Over time, this cycle of visitation, appreciation and action leads to new park creation, expanded influence of a park and further cycles of visitation, appreciation and action leading to more parks. Figure 1.8 shows the tourism and conservation cycle. Only when sufficiently large numbers of people in a society visit and value parks are sufficient public resources made available for establishment and management. Only when society has high levels of park demand can parks outcompete the other needs and interests of society.

Tourism is the fundamental element that determines whether a society has sufficient levels of conservation appreciation to lead to action. This appreciation must be consistent and ongoing. Other competing forces in society are always present and will subsume the land and the financial resources going to parks unless a mobilized public park constituency is always present and active.

Case Study Number 1: Madikwe Game Reserve (South Africa)

An Example of Ecological Restoration Designed for Tourism and Paid for by Tourism

Established in 1991, Madikwe is managed by the North-west Parks Board of South Africa. The reserve contains a restored African savannah ecosystem. Most of the reserve was once derelict farmland. Many derelict farm buildings and structures, hundreds of kilometres of old fencing and many alien plants were removed. Some buildings were spared and now serve as park offices and workshops, while various outposts have been built to house game scouts and other reserve staff. Approximately 60,000 ha of the reserve were enclosed by a perimeter fence measuring 150 km. This was later electrified to prevent the escape of elephants and the larger predators. Where possible, local business and labour were used to demolish and clear unwanted structures, erect fences, construct roads and build dams and lodges. Several game lodges have already been built. Other lodges will be developed in the future.

Wildlife reintroduction began early in 1991, shortly before the perimeter fence was completed. Operation Phoenix, as the reintroduction programme is called, is the largest game translocation exercise ever undertaken in the world. More than 10,000 animals of 28 species have so far been released into the reserve, including elephant, rhino, buffalo, lion, cheetah, cape hunting dog, spotted hyena, giraffe, zebra and many species of antelope and herbivores. Leopard already occurred in the reserve.

Madikwe functions through a system designed to benefit the three main stakeholders involved in the reserve. These are the North-west Parks Board, the private sector and the local communities. All three work together in a mutually beneficial 'partnership in conservation and tourism'. The North-west Parks Board is responsible for setting up the necessary infrastructure and the management to run Madikwe as a major protected conservation area in the North West Province. It also identifies suitable sites within the reserve, which are leased to the private sector for tourism-based developments and activities.

The private sector provides the necessary capital to build game lodges and to market and manage the lodges and the tourism and trophy hunting activities in the reserve. In this way, private sector money, rather than state funds, is used to develop the tourism potential of the reserve. By 1999, with only three of the ten planned lodges constructed, the economic impact of tourism was already larger than that of the farm operations that had been removed.

The Madikwe Wildlife Reserve was granted the British Airways/World Conservation Union award for Park Tourism in 1998. It is a superb example of ecological restoration, public/private cooperation and advanced ecotourism design (web site: www.parks-nw.co.za/madikwe/index.html).

References

Bly, R. (1980) *News of the Universe.* Sierra Club Books, San Francisco, California.

Burns, S.P. and Warren, C.L. (2000) *National Wildlife Areas and Migratory Bird Sanctuaries.* Homepage of the Canadian Wildlife Service, online. Available at: www.cws-scf.ec.gc.ca/hww-fap/nwambs/nwambs.html

Burton, P. (1986) *Vimy.* Penguin Books, Markam, Ontario.

Foster, J. (1978) *Working for Wildlife: the Beginning of Preservation in Canada.* University of Toronto Press, Toronto, Ontario.

Green, M.J.B. and Paine, J. (1997) State of the World's Protected Areas at the End of the Twentieth Century. Unpublished paper presented at *Protected Areas in the 21st Century: From Islands to Networks.* World Commission on Protected Areas, Albany, Western Australia.

Huth, H. (1957) *Nature and the American: Three Centuries of Changing Attitudes.* University of Nebraska Press, Lincoln, Nebraska.

Luke 4: 1, 2, 14, 15. King James Version of the Christian Bible.

Mabey, R. (1980) *The Common Ground: a Place for Nature in Britain's Future?* Hutchison, London.

Marty, S. (1984) *A Grand and Fabulous Notion: the First Century of Canada's Parks.* NC Press, Toronto, Ontario.

McFarland, E. (1982) The beginning of municipal park systems. In: Wall, G. and Marsh, J. (eds) *Recreational Land Use: Perspectives on its Evolution in Canada.* Carleton University Press, Ottawa, Ontario.

Michener, J. (1985) *Texas.* Random House, Toronto, Ontario.

Ontario Ministry of Natural Resources (2000) *Park Classification.* Homepage of Ontario Ministry of Natural Resources, online. Available at: www.mnr.gov.on.ca/MNR/parks/B3.html

Ontario Ministry of Natural Resources (2001) *Land Use Categories.* Homepage of Ontario's Living Legacy, online. Available at: www.ontarioslivinglegacy.com/protect2.html

Parks Australia (2001) *Uluru – Kata Tjuta National Park Plan of Management.*

Homepage of the Department of Environment and Heritage, online. Available at: www.biodiversity.environment.gov.au/protecte/pubs/plans/uluru_plan_download.htm

Parks Canada (2000) *National Historic Sites*. Homepage of Parks Canada, online. Available at: parkscanada.pch.gc.ca/nhs/english/nhsptxt_e.htm

Raffin, J. (1999) *Bark, Skin and Cedar: Exploring the Canoe in Canadian Experience*. Harper Collins, Toronto, Ontario.

Sax, J. (1980) *Mountains Without Handrails*. The University of Michigan Press, Ann Arbor, Michigan.

USFS (2001) *The Wilderness Act*. Homepage of the United States Forest Service, online. Available at: www.fs.fed.us/outernet/htnf/wildact.htm

USFWS (2001) *Conserving Wildlife and Habitats*. Homepage of the United States Fish and Wildlife Service, online. Available at: www.fws.gov/conwh.html

CHAPTER 2
Park Tourism in the World

Introduction

Parks and their use by people have a long history. Today's parks, recreational uses, facilities and values are the tip of the historical iceberg. What one sees today is much like what one sees when observing an iceberg in the ocean: the foundations are hidden in the past and are only visible to the discerning observer with careful thought and understanding.

This chapter provides an overview of the history of people's use of parks. It concentrates on the history of the peoples who share English as a common language and heritage.

An Overview of Park Tourism History

In Western society, the creation of parks and their use by people extends for over 500 years. Henry VIII, the King of England from 1509 to 1547, was a keen huntsman. He acquired hunting land on the outskirts of London because of the large number of deer and wild boar living in the area. This was the start of the famous Hyde Park that now graces central London.

Within 60 years, Charles I (reigned 1625–1649) opened Hyde Park to the public for the first time and had the ring road laid out, providing a drive for people to show off lavish horses and carriages. He also introduced 'Milk From a Red Cow', something of an early vending machine, where milk was provided straight from the cow in the park (Hyde Park, 2001). All classes of society used the park extensively for centuries for many purposes, recreational and social. The idea of a green park for public use in the centre of a city became entrenched in British society and was carried by government officials and emigrants throughout the British Empire.

In 1634, the Boston Common was set up by the British colonial authorities of Massachusetts. It began as a pasture where colonists could keep their cattle, a practice that continued until 1830 (Anon., 2001). The Boston Common is recognized as the first public park in the USA. It is the most famous of the New England 'commons', or common lands, around which New England towns were organized. It evolved into a city park and retained its character as green space in the heart of the city. The name 'common' is significant. This signifies that commoners, common people, had rights of access and use. Therefore, the first park formally established in the New World had inherent public use rights, a concept that continued and expanded in other towns and situations through the centuries that followed.

In 1763, the Halifax Common was granted to the City of Halifax by the Lieutenant Governor of the British Colony of Nova Scotia. It was first used as community pasture and for military exercises. The Halifax Common later became city parkland and is now located in the heart of the city. It is recognized as the first park in Canada.

These early North American city parks in Boston and Halifax mimicked London's Hyde Park model, a public green space open to the public in the centre of a growing city. As British North America split in the 1770s into the two countries of the USA and Canada, the British colonial parks of Boston and Halifax remained as progenitors in their respective countries.

In 1832 the US government took action indicating that a national government had a role in providing public recreation facilities. In that year the US Congress established Hot Springs Reservation to protect hot springs flowing from the south-western slope of Hot Springs Mountain in Arkansas. The springs and their hot water were used for therapeutic baths. Importantly, this reserve involved government protection of a natural resource for public health purposes, thereby establishing the American precedent of national government action for the reservation of sites for public recreational and health use. The site became a National Park in 1921. The Arkansas Hot Springs is the oldest park currently in the National Park System of the USA.

In 1832 George Catlin, an artist, proposed that that a 'nations park' be established in the USA for the protection of the plains Indians and the land and resources on which they depended. His ideas were that the national government should establish a combined ecological and cultural reserve and manage it for the protection of a culture and the landscape. The US government did not accept his idea. It took another 50 years before this idea of a 'nation's' park took root in the USA.

Through the 1840s to 1860s the rapidly developing cities in central Canada started to create public parks and squares. These included Toronto in 1848, Hamilton in 1852 and London in 1869, all cities in the British Colony of Upper Canada, later the Province of Ontario.

In 1851, Toronto City Council established the Committee on Public Walks and Gardens to manage public lands. The first job of the committee was to manage public recreational use of lands located in a federal military reserve, the Garrison Reserve. This committee is recognized as the first park management administration in Canada.

In 1864, the Mariposa Grove in the Yosemite Valley was established as a California State reserve by the United States Congress. The national legislation made it clear that the 'State shall accept this grant upon the express conditions that the premises shall be held for public use, resort, and recreation; shall be inalienable for all time' (Dilsave, 2000a). This legislation established the principle that the reserve cannot be sold and that it is for public use, resort and recreation. This grove became part of Yosemite National Park in 1890.

In 1865 Frederick Law Olmstead wrote the first management statement for this reserve in California. He stated eloquently that the contemplation of nature was good for human health:

> It is a scientific fact that the occasional contemplation of natural scenes of an impressive character, particularly if this contemplation occurs in connection with relief from ordinary cares, change of air and change of habits, is favourable to the health and vigor of men and especially to the health and vigor of their intellect beyond any other conditions which can be offered them, that it not only gives pleasure for the time being but increases the subsequent capacity for happiness and the means of securing happiness.
>
> (Olmstead, 1865)

This contemplative view of nature became a major driving force in national parks tourism. The idea that a park should have a plan directing development and operations started with Olmstead, but took over 100 years from this date to become widely accepted.

In 1866, the British Colony of New South Wales reserved 2000 ha (5000 acres) of land in the Blue Mountains west of Sydney for protection and tourism. This reserve protected the Jenolan Caves. Tourism was a key feature of this reservation, as indicated by the legislation which was intended to protect 'a source of delight and instruction to succeeding generations and excite the admiration of tourists from all corners of the world' (Hall, 2000). This was followed by the reservation of Jamieson Creek in 1870 and the Bugonia Lookout in 1872, also in the Blue Mountains. New South Wales was well on its way to establishing a reserves system, based on public ownership of important scenic areas. Public recreational use of these lands was a primary focus of the reserves. Blue Mountains National Park is now a well-used and recognized park. Strangely, the Australians have never made claim to having established the first national park, a claim they could legitimately make.

By Act of the US Congress on 1 March 1872, Yellowstone National Park was 'dedicated and set apart as a public park or pleasuring ground for the benefit and enjoyment of the people' and 'for the preservation, from injury or spoilation, of all timber, mineral deposits, natural curiosities, or wonders . . . and their retention in their natural condition' (Dilsave, 2000b; NPS, 2001a). The key concepts of a public park open to all for benefit and enjoyment became the cornerstone of national park establishment worldwide. Yellowstone is often recognized as the first national park in the world; however, the Arkansas Hot Springs and the Mariposa Grove in Yosemite also make similar claims.

The British Colony of New South Wales created a National Park in 1879. The park, subsequently known as Royal National Park, was created for public recreation. Given its location only 32 km south of the burgeoning city of Sydney, it was designed for city residents to escape from the pressures of urban living and enjoy themselves in a semi-natural setting (New South Wales Parks and Wildlife Service, 2000). In the early days this park was heavily modified to look like an English garden, a very difficult task in the Australian climate. Later more sensitivity to the native Australian flora and fauna developed and more ecologically appropriate landscape management occurred. This park is often recognized as the second national park in the world. The legislation creating this national park used the term national park, the first time any colony or country used this phrase in legislation (Hall, 2000).

In 1883, the Province of Ontario in Canada passed the Public Parks Act. This Act was the first Canadian legislation providing a legal structure for the establishment, general development and management of parks in municipalities in Canada. Later all other provinces would follow Ontario's lead in providing a legal structure for cities and towns to create and manage parks.

In 1885, the thermal hot springs in the Bow Valley of the Northwest Territories were protected as a reserve by regulation by the government of Canada. The westward expansion of the Canadian Pacific Railway had opened up this area to exploitation. The Canadian government decided to reserve the hot springs for public use, rather than let them be exploited by private interests. Clearly, the government was following the well-known lead of the US government in the reservation of hot springs for public use. This hot springs reserve was enlarged and became Rocky Mountains National Park in 1887. In 1930 it was renamed Banff National Park. Banff National Park is often recognized as the third national park in the world.

At the same time as the Banff hot springs protection debate was under way, the province of Ontario and New York State were engaged in discussions concerning the management of tourism at Niagara Falls, a shared resource. Ontario asked the Canadian national government to create a national park at Niagara, but this was refused due to political objections about the high cost. Ontario moved independently and in 1885 passed legislation that led to the creation of Queen Victoria Niagara Falls Park. The New York State Reservation at Niagara Falls also became a reality in 1885 (Seibel, 1995). The overall goals of the parks were to provide public access to Niagara Falls on publicly owned and managed land. In Canada, this involved the removal of private development from the edges of the Niagara Falls and Gorge, and its replacement with public parkland. In both countries special purpose park and tourism management agencies were created. The idea that the public sector could do a better job than the private sector in the administration of public recreation in special natural sites was reinforced with this Ontario and New York decision and became a fundamental component of park management in North America. In addition, the international cooperation in park planning and management, started at Niagara Falls, continued and expanded in the coming years. The creation of parkland by Ontario and New York State set a precedent that provinces and states, not just national governments, had a significant role in park, tourism and recreation management. Subsequently, all provinces in Canada and states in the USA created their own park systems.

In 1885, the state of New York in the USA created the Adirondack Forest Reserve, with the goal of preserving important headwaters and creating a 'pleasuring ground' for the expanding population of the state. With legislative strengthening in 1892 and 1894, the concept of 'forever wild' was introduced. This concept stated that wild nature, in this case the deciduous forests, lakes and mountains, has inherent value in its own primeval state. This is a widespread concept now, but was revolutionary in 1892. This area is now a state park. Significantly, the majority of the land is still in private ownership and there are stringent development laws governing this private land. The idea of a greenland park, comprising mixed public and private land, did not become popular elsewhere, except in the UK in 1949 with the passage of the National Parks and Access to the Countryside Act. This Act allowed for national park designation in England and Wales on private land.

In 1887, Last Mountain Lake in Canada's Northwest Territories (now Saskatchewan), was designated as Canada's first federal wildlife sanctuary. Important areas for nesting and feeding of migratory waterfowl were protected from settlement and agricultural development spurred by railway development in the area (Foster, 1978; Environment Canada, 2001). This was the start of the National Wildlife Area system in Canada, but it was not until 100 years later that the area was officially designated as a National Wildlife Area. This is the first national designation for the protection of wildlife and their habitat in Canada. This sanctuary kept its primary purpose of protection and did not develop or encourage visitation. As a result, visitation and public profile stayed low. Similarly, the National Wildlife Areas of Canada never developed a significant public constituency and today have only a tiny public profile and a small budget compared with National Parks. The development of visitation and public profile became a critical element in the development of a public constituency for a park system. Those systems with a mobilized public constituency were able to grow and prosper, those that did not languished.

Clearly, in the 1880s, national governments, states and provinces in Canada and the USA were experimenting with various forms of protected areas. These gave rise to national parks, forest reserves, wildlife sanctuaries, state parks and provincial parks.

On 1 April 1890, the US Congress set aside Chickamauga and Chattanooga Military Park as a national military park. These US Civil War sites became the first of many national historic sites to be established in the USA. The initial goal of these sites was historical and professional military study. The national historic parks and sites systems of the USA developed into important benchmarks and interpreters of the development of the country, its peoples and cultures. The National Park System of the USA now represents 56 different historical and cultural themes within its 378 units (NPS, 2001c). These sites are an important element in the protection and interpretation of the culture of this country.

In 1893, the government of Ontario created Algonquin National Park, the second major park created by a province or state in North America. This was a huge area of highland forests and rivers. In the early years its primary goal was to establish sustainable forest harvesting and to protect important headwaters. Later, the park was renamed a provincial park and was developed for wilderness recreation. Algonquin Park became the model for provincial parks in Canada as it developed a substantial public constituency in the urban populace of Canada's most populated province, while retaining significant natural and recreational values.

In 1887 in New Zealand, sacred mountains and the surrounding lands were given by the local Maori people to the crown. In 1894, New Zealand created Tongariro National Park around these sacred mountains (Booth and Simmons, 2000). This intimate connection between an aboriginal people and a colonial people in the creation of a park unfortunately remained very rare until 100 years after the New Zealand innovation.

In 1898, The Volksraad (government) of the Zuid-Afrikaansche Republiek, under the leadership of President Paul Kruger, proclaimed the Sabi Reserve. In 1903, the Shingwedzi Game Reserve was proclaimed. On 31 May 1926, the Sabi and Shingwedzi reserves were formally united and proclaimed as Kruger National Park. The National Parks Board (now South African National Parks) was established in the same year, within the terms of the National Parks Act passed by Parliament (Kruger National Park, 1998). Kruger National Park developed a significant international profile over time and is now the anchor of the nature-tourism industry in South Africa.

In 1903, President Theodore Roosevelt of the USA set aside a tiny island off the east coast of Florida for the protection of pelicans and other species of birds from market hunters. Five acres, known as Pelican Island, started what is now known as the National Wildlife Refuge System of the USA (USFWS, 2001). This action was the start in the USA of the creation of wildlife reserves for the protection of wildlife and the management of sustainable harvests. This system was closely tied to various user groups; as a result it developed a reasonably strong public profile.

In 1906, the State of Queensland in Australia passed the State Forests and National Parks Act. This was the first legislation in the world providing the procedures to be followed in establishing national parks (Hall, 2000). Before this, each park was created by special purpose legislation and with no concept of an organized system of parks.

In 1906, the US Congress passed the American Antiquities Act. This legislation gave the President power to 'declare by public proclamation historic landmarks, historic and prehistoric structures, and other objects of historic or scientific interest that are situated upon the lands owned or controlled by the Government of the United States to be national monuments'. The Act also provided for administration of the historic monuments (NPS, 2001d). This was the first federal Act in the USA providing for protection of archaeological sites and for protection of prehistoric and historic sites on federal lands. This legislation laid the groundwork for the national historic parks and sites system that developed in the USA. The Act provision giving the President the power to declare national monuments became a powerful tool for protected area establishment in the USA. Many national monuments created by the President were later declared national parks by legislation passed by Congress.

In 1911, the Parliament of Canada passed the Dominion Forest Reserves and Parks Act. This provided a coherent structure for the management of forest reserves and national parks and a government agency. All forest reserves and parks were managed under one piece of legislation and one agency. This was the first national park legislation that provided procedures for park management. The legislation also enabled the creation of the Dominion Parks Bureau, first headed by James Harkin. This was the first national park agency in the world and Harkin the first director. Under the leadership of Harkin, the Parks Bureau was successful in seeing the number of national parks increase to 18 by 1930 (Marty, 1984). Harkin was a visionary who worked to develop a national parks constituency in the country and in Parliament. He developed tourism management regimes that allowed use, but with controls over negative impacts. His work established an agency that recognized that public use was essential if citizens were to understand and support national parks. The agency went on to develop a very high public profile and the parliament responded with comparatively high levels of budget.

In 1916, the US Congress created the National Park Service, following Canada's lead, and passed organic legislation to govern all national parks in the country. The legislation stated that:

> The service thus established shall promote and regulate the use of the Federal areas known as national parks, monuments, and reservations hereinafter specified by such means and measures as conform to the fundamental purpose of the said parks, monuments, and reservations, which purpose is to conserve the scenery and the natural and historic objects and the wild life therein and to provide for the enjoyment of the same in such manner and by such means as will leave them unimpaired for the enjoyment of future generations.
>
> (Dilsave, 2000c)

This important wording in the legislation, 'unimpaired for the enjoyment of future generations', dictated that national parks are for public enjoyment, with the important caveat that such use shall leave them unimpaired. This wording set the tone for national park use for the next century: public use with little negative environmental impact.

After local citizens lobbied to have the site preserved, the government of Canada started to administer Fort Anne in Nova Scotia as a National Historic Site in 1917. As the focal point for French and British settlement and as the seat of government of the French Acadia and later the English Nova Scotia, Fort Anne played an important role in Canadian history. The site was the scene of numerous battles as France and England fought for control of North America in the 17th and 18th centuries. Fort Anne became the first of many sites protected, restored and commemorated by the national government. Later the national historic parks and sites were combined with national parks under the administrative umbrella of one agency, now called Parks Canada. Both the USA and Canada utilized one national parks agency for the management of both historic and natural national parks. Over time, the commemoration of important historic sites within a national historic park system became a valuable element of American and Canadian culture.

In 1930, the Canadian Parliament passed the National Parks Act. This was Canada's revised parks act governing national parks. It used some of the key concepts found in the US National Park Service Organic Act, such as the concept of protection for all time with compatible public recreational use.

In 1935, the USA Historic Sites Act was established. This Act 'declared that it is a national policy to preserve for public use historic sites, buildings, and objects of national significance for the inspiration and benefit of the people of the United States' (NPS, 2001e). This was important legislation in the development of the national historic park and site system in the USA.

On 16 December 1946, Nairobi National Park was gazetted by the British colonial authorities in Kenya. This was the first national park to be

Fig. 2.1. Tsavo National Park is a prominent Kenyan Park, heavily used by international tourists. Tsavo's prominent gateway emphasizes to all visitors that one is entering a special place. Tsavo National Park gate, Kenya. (Photographed by Paul E.J. Eagles.)

created in Eastern Africa (Kenya Wildlife Service, 2001). It was the first of many parks and game reserves in this part of Africa. These reserves became the backbone of the Eastern Africa tourism industry.

In 1949, the British Parliament passed the National Parks and Access to the Countryside Act. This provided the legal basis for the creation of National Parks, National Nature Reserves (NNRs), Local Nature Reserves (LNRs) and Sites of Special Scientific Interest (SSSIs) (JNCC, 2001). The British approach deviated considerably from that pioneered in North America, in that the creation of the British protected areas often occurred using the regulatory powers of government over private land, rather than government land ownership. Therefore, many of the sites remain in private ownership with regulations governing the types of activities allowed. This type of approach is sometimes called greenfield planning.

This brief history covers 500 years of parks across the English-speaking world. Throughout this history of park development, several obvious themes occur. All the parks were created with public benefit in mind. Health benefits, physical, mental and spiritual, were paramount. Public use was inherent. Once created, public satisfaction typically assured their continuance. Over time, it was recognized that human use could be destructive of park values and that protection measures were needed.

In North America the very first reserves were for military purposes (protection of forests for naval construction purposes, areas for military training) or for common use purposes, such as grazing and recreation. Both of these types of reserves often later became parks with conservation and recreation as goals.

All levels of government – cities, colonies, provinces and states, and federal governments – created parks. In due course, the sophistication of management increased. Agencies were created to provide professional management, first in cities, later in nations and later still in provinces and states.

Historic Themes

The European countries had the high culture of the theatre, literature, art, architecture and the church, while in the early years their colonies lacked all of this. The European colonialists to North America, New Zealand, Australia and South Africa were constantly reminded of the superiority of the obvious icons of the Old World culture. However, the wild lands and wildlife of the New World were widely available and were not generally present in Europe. They slowly evolved as a cultural icon to help to define national life; over time the concept of a national park, and other similar parks, helped to coalesce these ideas into a cultural icon.

National parks and wildlife refuges, using the North American model, contained several key concepts, including:

1. Wild, abundant nature without permanent human habitation;
2. Nature largely free from human interference;
3. Common property;
4. Public administration; and
5. Available to all people for recreation.

Initially, the parks were individually created. Later, with burgeoning park numbers, a coordinated management structure across a system of parks developed.

After a while, the parks diverged into different categories, such as city parks, wildlife reserves, state/provincial parks and national parks. Each developed a culture, a type of use and a public constituency. The systems often became administratively isolated from each other. Generally park visitors were less likely to adopt this segregation. Most outdoor recreation users visited many different types of parks.

It is important to note that the creation of parks without human residents was largely dependent on governments extinguishing the land rights and uses of aboriginal peoples. Once these people were removed and their activities eliminated, it could be stated that the area was wilderness, that is, an area wild and without human activities. The treatment of aboriginal land rights took very different tacks in different countries, and is too complicated to deal with properly in this book. However, it must be remembered that before the lands in the New World were occupied by European colonists and their culture, they were used to various degrees by native peoples.

Strangely, it was not until the later half of the 20th century that colleges and universities started to produce park managers with training in the various fields required for park planning and administration. Initially, the training often concentrated on natural resources, with less emphasis on the critical aspect of human and visitor management. The early leadership role in the development of university programmes in park, leisure and recreation management occurred in the USA, with most other countries following later, using the American programmes as models.

Throughout these five centuries, the public use of parkland was the spark-plug keeping the park movement alive. This use created further demand. This demand created a management response, in the form of keepers, guards, guides and agencies. It also prompted governments to create more parks. As soon as numbers became prominent, a business lobby developed aimed at exploiting the tourism flow. For centuries, the needs of the park, of the individuals visiting and of the businesses servicing the tourists were in a constant tussle. Each is necessary, but their needs often conflicted. The success of park management is often measured by the balance between individual and group interests in park tourism.

The State of Park Tourism (Volume and Distribution)

As parks were created, they were used for a wide variety of recreation. The tabulation of the amount of use was sporadic and often unreliable. However, in most competent park systems, procedures were developed for the recording of volumes and types of uses in the parks. Such use data were provided by the park agencies to governments and to interested stakeholders. Today, any student of parks can find reasonably accurate data on the amount of visitation occurring in the various developed-world park systems.

No international inventory of park tourism has been made. Each park jurisdiction keeps its own records, with little effort to create overall tabulations. Such jurisdictional records can be used to gain an understanding of changes over time, but only for individual park systems.

Agency data typically show the increasing popularity of parks over time. In Canada, Parks Canada has seen use levels increase from a handful in 1885 to 52,000,000 visitor days per year by 1998. Ontario Provincial Parks steadily increased from a handful in 1893 to 10,000,000 visitor days of recreation in 2001. The US National Park Service increased from a few visitors in 1872 to an incredible 287,130,879 in 1999 (NPS, 2001b).

The recreational use of ten UK national parks reaches up to 76 million visitors (Parker and Ravenscroft, 2000). However, it is important to note that this number is difficult to estimate due to the open nature of these parks with multiple entrance points, few control gates and large populations living in the parks.

Kenya National Parks has increased from 4000 visitors in 1950 to around 160,000 a year now (KWS, 2001). This number is now in danger of decreasing due to the civil disorder in Kenyan society that is inhibiting tourism.

The National Parks of Costa Rica saw visitation rates increase from 250,000 in 1985 to 860,000 in 1999 (Baez, 2001). This large increase corresponded to the development of ecotourism as a major economic activity for this small Central American country.

These changes observable in the USA, Canada, Kenya, Costa Rica and the UK show general trends. Over time, more parks are created and more visitation occurs. It is clear that these two are related. As more visitation occurs, more people become supportive of parks and public demand for more parks strengthens. As more parks are created more opportunity for tourism becomes available. This supportive cycle of creation and visitation is a standard international trend.

These use trends demonstrate that these sites produce important public benefits. They also suggest that there are important consequences flowing from this use. These include social, economic and environmental impacts; some considered negative, others positive and some neutral.

Government policy has a significant impact on park tourism. Tanzania saw a large drop in tourism usage when government policy closed the border with Kenya. The goal was to force international tourists to fly directly to Tanzania and thereby bypass Kenya. The policy was a failure, as the foreign tourists did not shift their arrival point from Kenya to Tanzania. The foreign tourism levels dropped from 250,000 in 1973 to slightly over 50,000 in 1983. Once government policy change in the early 1990s again allowed cross-border transit, the country saw an increase to 318,000 by 2000 (Wade *et al.*, 2001). These national tourism changes were mirrored in park usage.

Kenya saw a significant increase in park visitation through the 1980s and 1990s. In the later 1990s, widely reported civil disorder in the country resulted in a significant drop in foreign travel to the country. Park use dropped accordingly. This is a major problem for Kenyan parks because the Kenya Wildlife Service is funded from tourism fees and charges. Lower levels of tourism income means lower levels of budget available for park management.

In Canada and the USA, an overall tabulation of park visitation in state/provincial and national parks found an estimated 2,626,275,241 visitor days of recreation activity for 1996 (Eagles *et al.*, 2000). They calculated an associated economic impact of between US$236 billion and 370 billion for that year. These very impressive figures reveal the importance of park visitation to the people of these countries and to their economies.

The very high figure does not include regional or city park use levels. The authors of the work also stated that this use level is an underestimate for the agencies reported, due to structural and financial limitations within the park agencies. Clearly, park visitation and park tourism is a massive phenomenon in these two countries.

The World Conservation Monitoring Centre, a division of the United Nations Environment Programme, is collecting a global inventory of park use. This first international tabulation of park tourism will be very useful when it becomes available.

Park tourism is a massive and growing cultural, social and economic phenomenon. As special natural and cultural lands become scarcer with increasing development and expanding human populations, these lands will become even more precious. Therefore, increasing demand is expected.

The Goals and Impacts of Park Tourism (Social, Cultural, Economic, Ecological)

All government actions create an impact; that is their purpose. Governments make policies in order to create a desired outcome. The creation and management of a park has long-term impacts. Other possible land use options are forgone. Some existing land uses are eliminated, others accentuated.

It is important to recognize that the determination of whether an impact is positive or negative is highly subjective and situational. Any one impact may be highly positive to one person but highly negative to another. Any one impact may be both positive and negative for the same person in different aspects of their life. In this section of the book, the authors try to refrain from labelling categories of impacts as positive or negative. Such a determination must be made by those affected.

The creation of a park is usually designed to produce a desirable social impact; the creation of a satisfactory leisure and cultural condition. Eagles *et al.* (2002) propose that park benefits can be seen to accrue at three levels: society, park and tourism management, and individual visitors. Society creates parks for the following social benefits:

- redistribute income and wealth;
- increase opportunities for employment;
- gain foreign currency;
- assist community development;
- promote the conservation of natural and cultural heritage;
- sustain and commemorate cultural identity;
- provide education opportunities to members of society;
- promote health benefits; and
- expand global understanding, awareness and appreciation.

These benefits are those that accrue to society at large. They are the types of benefits desirable to national or regional levels of society and government. However, those directly involved in tourism management

Fig. 2.2. Creation of a park is usually designed to produce a desirable social impact. One goal at Louisbourg National Historic Site is to restore and interpret an important period of French–English conflict in Canadian history. Another goal is to stimulate the tourism industry in a remote corner of Cape Breton Island and thereby provide valuable jobs and economic benefit. The work on the Louisbourg French Settlement was the largest and most expensive historic restoration in Canadian history. Fortress of Louisbourg National Historic Site, Canada. (Photographed by Paul F.J. Eagles.)

seek a different, more targeted range of social benefits. Tourism operators and park managers view tourism as a means to:

- promote conservation;
- develop heritage appreciation;
- generate revenue;
- learn from others;
- create employment and income;
- develop long-term sustainable economic activity;
- make a profit;
- manage resource extraction;

- foster research; and
- create a positive experience.

Individuals seek out parks for personal benefits, or for the benefit of family members or friends. It is these individual searches that provide the basis for park travel, for park tourism and ultimately for societal justification of parks. Individuals seek out experiences in parks to:

- promote conservation and preservation;
- gain health benefits;
- enhance personal experiences, which include:
  - cognitive objectives (for example, learn about nature and wildlife);
  - affective concepts (for example, gain peace of mind);
  - psychomotor desires (for example, get exercise);
- participate in a social experience;
- achieve family bonding;
- spend quality time with peers;
- provide the opportunity for courtship rituals;
- meet people with similar interests;
- achieve group team building;
- achieve time and cost efficiency;
- feel personal accomplishment;
- explore history; and
- reaffirm cultural values.

The attainment of these benefits has cultural impacts. In the fullness of time, societal shifts take place as parks, their activities and impacts are incorporated into the social fabric. The older the park the more likely it will be a vital cultural feature of a community. Hyde Park is as much a cultural element of London as is the central business area or the theatre district. The flow of people to and from a park creates cultural impacts on the people who live around the park. People like to live close to a park. Tourism businesses providing food, entertainment or souvenirs are located in primary locales of heavy traffic. Esoteric activities, such as sexual services, drama schools, retirement homes and photography studies, have been known to locate near parks. Over time literature, drama and film develop cultural images and symbols for a park.

Travel moves people and money around. The economic impact of a park can be substantial, but is often only weakly comprehended during designation and management. Residential land values are often very high near parks. The spending of people in direct recreation participation is an economic diversion from another place and activity. A park often creates demand for the clothing, equipment and supplies that people need to properly utilize the experiences available. Sometimes these leisure demands create cottage industries of importance. Many parks have substantial cultural industries comprising film, art and craft creations. At very high use levels, services are required, such as high-volume roads, sewage and water systems, electrical services, policing services and all the trappings necessary to service urban populations. The financial expenditures of such infrastructure can rival those of city development.

Parks have major environmental impact. The park will have a certain environmental identity, such as a country garden or a wilderness. The fulfilment of this identity has impacts on all aspects of the natural environment, ranging from wildlife, through plants to soils. Over time, park scholars and managers have produced a large literature on the environmental impact of outdoor recreation. This literature is used in the design of parks and in their management.

Case Study Number 2: Soufrière Marine Management Area (St Lucia)

Tourism Management in a Sensitive Marine Environment

Soufrière Marine Management Area was established in 1994 to protect and manage marine resources, including coral reefs, near the island of St Lucia in the Caribbean Sea. It is funded by various government agencies and by the sale of goods, user fees and an active 'Friend's Group' of volunteers. The park is managed by the Soufrière Foundation and Department of Fisheries, under the guidance of a Technical Advisory Committee comprising key management authorities and user groups. The park attracts around 3600 yachts and 5000 boats a year, with 21,000 snorkellers and 12,000 divers.

The park was created after a lengthy negotiation process between the government, the local fishing community, dive operators and other shore-based interests. The park tourism replaced some of the local reliance on fishing. The fishermen gave up access to near-shore fishing areas in return for other employment opportunities, provision of small business loans and other economic alternatives. The marine reserve now provides excellent fisheries, spawning and replenishment functions for a much larger area.

The park area is divided into zones, each with an activity profile. Some polluting activities are placed into zones acknowledging the activity. The park managers have established a visitor management programme, with specific emphasis on controlling the numbers and activities of the divers attracted to the marine resources. Yacht anchoring is restricted to a few selected areas. In these yacht zones, approximately 60 mooring systems have been installed for anchoring. Marine biologists have confirmed that, due to these management actions, some reefs are already showing progress in recovering from previous damage.

The park has an active education and interpretation programme. Lectures, brochures, video and the Internet are used to disseminate information on the park and its use. Private dive operators are required to give specific training and guidance to divers.

The Soufrière Marine Management Area has monitoring programmes designed to measure the change of the ecosystem in response to human activities and environmental enforcement. The routine measurements attempt to track environmental, biological and socio-economic variables at key locations in the area over the long term. Research projects are designed to improve the knowledge and understanding of the structure and function of the marine ecosystem of the Soufrière coast. They range from studies of coral fish migration to profiles of the people in the Soufrière community.

Soufrière Marine Management Area of St Lucia was granted the 1997 British Airway/ World Conservation Union Award for excellence in park tourism. It is a good example of careful community development, multi-stakeholder management, sensitive tourism development and good planning. The management frameworks allow visitor use without negative impact on the marine resource. As a result of all the efforts an important marine resource in the Caribbean Sea is protected and well managed (website: www.smma.org.lc/).

The Need for Tourism Management

The management of park visitation is a fundamental component of park management. The size and scale of the impacts are such that sophisticated management systems are required. Chapters 4, 5 and 6 in this book provide details for planning and management of park tourism.

Trends Affecting Park Tourism

Demographics, economics, technology and changing lifestyles all play important roles in shaping the future of outdoor recreation and park management (Gartner and Lime, 2000). Several trends have a significant effect on park management; these are discussed and their implications highlighted. This discussion should allow managers the opportunity to evaluate their options for planning and reacting to forces that are changing the marketplace and the nature of protected area management.

As long as energy remains comparatively inexpensive, expansion in global tourism will continue. Increasing numbers of people visiting remote parks can be expected.

Demographic changes

The population structure in the developed world is changing significantly as the baby-boom generation ages. The generation is composed of those born between 1946 and 1964. Due to the large numbers in this group, it will have major impacts on the outdoor recreation market for many years. The ageing population has important implications for parks. Since people over the age of 45 camp much less frequently than those under the age of 45, this large population of older people are not likely to be involved in this activity. In other words, if seniors cannot be encouraged to camp or visit parks more frequently, the park market and outdoor recreation market in general is likely to decline significantly in years to come. On the positive side, older people tend to have both more time and more money at their disposal; the challenge is to encourage them to spend some of each in parks. It seems reasonable to suggest that the baby boomers could be encouraged to continue camping and visiting parks into their older years if the parks can maintain their interests and meet their needs as they age.

The baby boomers are currently among the most frequent campers, and if park managers can appeal to this generation's changing interests (with age), then the market for parks could remain strong into the future.

Older age groups prefer accommodation offering greater comfort. This will lead to the increased popularity of recreational vehicles and roofed accommodation in the future. While perhaps not realistic or appropriate for some parks, roofed accommodation is attractive to many. The activities that ageing visitors prefer will change in many cases; highly physical activities in most cases are abandoned in favour of birdwatching, wildlife viewing, pleasure walking, picnicking and related, less strenuous, pursuits. Such activities would seem to be well suited to a park environment, giving parks the ability to cater to this group by providing abundant and accessible opportunities for these activities to occur in the parks. It is important to keep in mind that the activities preferred by the baby-boom generation will probably be growth activities (Foot, 1996). The promotion of such activities does not necessarily have to be coupled with park camping; encouraging day visits for the older market may be deemed more appropriate or feasible if the desired accommodation cannot be provided. Alternatively, accommodation in nearby towns or villages could be promoted, or arranged as part of a package tour to the park involving several days of activities and programmes.

Throughout the world, populations of people are moving around at unprecedented levels. Some of these are permanent, such as emigration to new countries. Others are temporary, such as leisure and business travel. These new citizens and visitors commonly have very different experiences with outdoor recreation than they may have had traditionally. This has important implications for information provision in the parks. There needs to be a greater emphasis on basic information, park locations, rules, advanced reservations and recreational opportunities. Training in basic outdoor recreation skills will be required. Cross-cultural programming will need to be more frequent in parks. Providing a mix of services, facilities and opportunities that pleases a diverse group of park users will be a challenge.

Ecotourism

Ecotourists are those people 'interested in experiencing and learning about wild nature within natural settings' (Eagles, 1995). Ecotourism is considered to be the fastest growing of all tourism submarkets with tremendous potential for increasing visitation in parks' protected natural landscapes. Parks provide excellent opportunities for ecotourists, with their outstanding scenery, wildlife, wilderness, beautiful lakes, rivers, forests and beaches. Ecotourism benefits include the ability to educate visitors while supporting natural areas and resources through revenue generation. At the same time, impacts due to visitor numbers and

Fig. 2.3. The private game reserves on the fringes of Kruger National Park provide high quality ecotourism experiences and contribute to the conservation of the greater ecosystem outside the park. They also provide very valuable employment for local people. Elephant viewing at the Sabi Sabi Private Game Reserve, South Africa. (Photographed by Paul F.J. Eagles.)

activities must be carefully monitored (Eagles, 1995). While any one park system has significant potential to be a major ecotourist destination, this capacity has often not been fully realized; several obstacles and challenges face the development of an ecotourism/nature-based travel market to some parks. There is no shortage of destinations in other countries for ecotourists to choose from. If any one park system wishes to capture a portion of this rapidly expanding market, steps must be taken to overcome the challenges facing the development of this market globally.

Changing lifestyles and environmental awareness

Increasingly, many people strive to live an active lifestyle; a lifestyle that promotes 'wellness' or health of the whole being. For many, achieving this lifestyle involves reaffirming a connection with nature. Many parks have the opportunity to capitalize on this trend by marketing provincial parks as ideal spots to relax, breathe fresh air and exercise, while enjoying favourite recreational activities.

This change in lifestyle is closely related to a general increase in awareness of environmental issues among developed world societies. Since 1970, the year of the first 'Earth Day', environmental awareness has skyrocketed. Children are increasingly being exposed to the basics of ecology, waste recycling, environmental ethics, consumption and sustainability. It is these children who are developing into the ecotourists of tomorrow. Parks can contribute to furthering this education and provide a lasting impression by offering programmes for both children and adults that focus on environmental issues.

Technological advancement

The rate at which information exchange and access is now possible is changing the way we communicate and learn. It is also creating a world in which there are fewer and fewer knowledge barriers; the global community is shrinking and moving closer to everyone's doorstep. Now, people from the far corners of the world can more easily learn about and visit any one park. Some people are making international trips to visit these parklands, and they may or may not have the requisite knowledge and equipment to participate. They may require special amenities or programmes. The implications for communications, marketing, programming, accessibility and facilities are significant.

In the very near future, recreationists will have access to massive online databases containing topographical maps, wildlife locations and sounds, evaluative rankings of sites and much more. With the onset of digital communications linked worldwide through satellites and small, powerful, hand-held computers, each park visitor can have access to this data at any location throughout any park. With geographical positioning systems, each visitor can know their location precisely. When this geographical knowledge is connected to remote databases, the information component of outdoor recreation will be significantly enhanced. It is possible that the recreationist will have information access as good as or better than the park managers.

Furthermore, the rate at which technology is advancing and being improved means that there are more options for service delivery and for streamlining management efficiency. For example, a single personal computer can keep track of huge databases of campers, which can aid managers in making reservations and monitoring usage levels. With the right equipment, managers have the opportunity to transfer information efficiently and regularly between parks and head offices. The Internet is a powerful tool with which a park can market its products all over the world. If the application can be found and properly understood, the opportunities afforded by technology are practically boundless.

Other important global themes

The global park and tourism industry is massive. However, and rather unfortunately, most countries' parks systems are extremely undervalued in terms of their contribution to local and national economies. The result has been the conclusion that most parks and protected areas are not financially viable or economically sustainable. This is probably responsible for the significant budget cuts that many agencies suffered in the last decade of the 20th century. A much greater emphasis needs to be placed on carefully measuring and documenting the positive economic and non-monetary impacts of parks. Parks agencies will increasingly have to approach protected areas management as a business if they are to receive the attention and funding they deserve from governments. This will require a stronger degree of professionalism and business training on the part of park managers.

Emerging out of the move towards greater professionalism have been two very important movements in the protected areas field. The first is international efforts to develop best practice guidelines and standards for visitor data collection and management. Recalling that managers must satisfy a large set of information needs to effectively manage visitors, the importance of how agencies and managers go about collecting and managing visitor data becomes evident. Standard procedures for calculating visitation are being adopted to ensure consistency and accuracy in databases and for making meaningful, interagency, statistical comparisons. Standardizing definitions of data collection terminology, 'visitor days' for example, is also on the agenda.

The second movement is an initiative to create and implement a practical framework for evaluating management effectiveness in parks and protected areas. Evaluating the success (or failure) of different services, programmes and management strategies is receiving a great deal of emphasis in many parks agencies.

The switch to a business model approach for park management results in a greater emphasis on user pay revenue generation strategies. This now places a renewed importance on park visitors as customers. In order to successfully satisfy its customers, managers must have a thorough understanding of the visitors, the markets and the macro-environment forces that shape each of them. Population shifts, growth in nature-based tourism, changing lifestyles, increasing environmental awareness and technological advancements are some of the key trends that will shape the future of outdoor recreation and protected areas management. This discussion is intended to highlight some of the implications but, more importantly, to encourage managers to further their understanding of these, and other, important trends affecting protected area management. The degree to which managers understand these forces, plan for and adapt to them will significantly influence the parks' ability to continue to provide high-quality experiences and generate revenue, while protecting precious natural and cultural environments.

Summary

Parks are created by governments through the stimulus of influential people. As the public visitation levels build up over time, more people in society develop an appreciation of these sites. This appreciation leads to political pressure for more parks and leads to demand for more visitation. This circle of visitation, appreciation and creation has become a self-reinforcing phenomenon in most countries in the world.

Park visitation/tourism is a fundamental element of the park phenomenon. Its development over the last 500 years is a laudable element of global culture.

The size of the global park tourism industry is very large and growing. As long as global energy remains inexpensive and widely available, this trend will continue.

Fig. 2.4. This home in Cavendish, Prince Edward Island (PEI), is famous as the setting for Lucy Maud Montgomery's classic tale of fiction, *Anne of Green Gables*. Anne's life personifies rural life in Prince Edward Island in the early 20th century. Green Gables in PEI National Park represents the period that was beautifully captured by the *Anne* series of books. As a historic site, it is unique in representing a lifestyle and a period, illustrated through the life of a fictional person. Literature can be a powerful definer of culture and values. Green Gables is an example of a strong link between a local culture, a landscape and community values. Green Gables House in Prince Edward Island National Park, Canada. (Photographed by Paul F.J. Eagles.)

References

Anon. (2001) *Boston Common*, online. Available at: sterling.holycross.edu/departments/english/sluria/bcommon.htm

Baez, A.L. (2001) Costa Rica Como Destino Turístico. Unpublished Conference Paper, Rio De Janerio, Brazil.

Booth, K.A. and Simmons, D.G. (2000) Tourism and the establishment of national parks in New Zealand. In: Butler, R.W. and Boyd, S.W. (eds) *Tourism and National Parks: Issues and Implications*. John Wiley & Sons, Chichester, UK, pp. 39–49.

Dilsave, L. (2000a) *An Act Authorizing a Grant to the State of California of the 'Yo-semite Valley', and of the Land Embracing the 'Mariposa Big Tree Grove'*, online. National Park Service, available at: www.cr.nps. gov/history/online_books/anps/anps_1a.htm

Dilsave, L. (2000b) *An Act to Set Apart a Certain Tract of Land Lying Near the Headwaters of the Yellowstone River as a Public Park*, online. National Park Service, available at: www.cr.nps.gov/history/online_books/anps/anps_1c.htm

Dilsave, L. (2000c) *An Act to Establish a National Park Service, and for other Purposes*, online. National Park Service, available at: www.cr.nps.gov/history/online_books/anps/anps_1i.htm

Eagles, P.F.J. (1995) Understanding the market for sustainable tourism. In: McCool, S.F. and Watson, A.E. (eds) *Linking Tourism, the Environment and Sustainability*. US Department of Agriculture, Forest Service, Intermountain Research Station, Gen. Tech. Rep. INT-GTR-323, Ogden, Utah, pp. 25–33 (republished at: www.ecotourism.org/datafr.html).

Eagles, P.F.J., McLean, D. and Stabler, M.J. (2000) Estimating the tourism volume and value in parks and protected areas in Canada and the USA. *George Wright Forum* 17(3), 62–76.

Eagles, P.F.J., McCool, S.F. and Haynes, C. (2002) *Sustainable Tourism in Protected Areas: Guidelines for Planning and Management*. IUCN, Gland, Switzerland (in press).

Environment Canada (2001) *Looking Back*. Homepage of Environment Canada, online. Available at: bertjr.mb.ec.gc.ca/nature/whp/lml/df09s03.en.html

Foot, D. (1996) *Boom, Bust and Echo: How to Profit from the Coming Demographic Shift*. Macfarlane Walter & Ross, Toronto, Ontario.

Foster, J. (1978) *Working for Wildlife: the Beginning of Preservation in Canada*. University of Toronto Press, Toronto, Ontario.

Gartner, W.C. and Lime, D.W. (2000) *Trends in Outdoor Recreation, Leisure and Tourism*. CAB International, Wallingford, UK.

Hall, C.M. (2000) Tourism and the establishment of national parks in Australia. In: Butler, R.W. and Boyd, S.W. (eds) *Tourism and National Parks: Issues and Implications*. John Wiley & Sons, Chichester, UK, pp. 29–38.

Hyde Park (2001) *History*. Homepage of Royal Parks, online. Available at: www.royalparks.co.uk/hyde/hismain02.html

JNCC (2001) *Nature Conservation in the UK: an Introduction to the Statutory Framework*. Homepage of Joint Nature Conservation Committee, online. Available at: www.jncc.gov.uk/communications/natcons/default.htm#top

Kenya Wildlife Service (2001) Homepage of Kenya Wildlife Service, online. Available at: www.kenya-wildlife-service.org/naipark.htm

Kruger National Park (1998) *Establishment*. Homepage of Kruger National Park, online. Available at: www.parks-sa.co.za/knp/centenary/establishment.htm

Marty, S. (1984) *A Grand and Fabulous Notion: the First Century of Canada's Parks*. NC Press Limited, Toronto, Ontario.

NPS (National Park Service) (2001a) *In Brief*. Homepage of Yellowstone National Park, online. Available at: www.nps.gov/yell/

NPS (National Park Service) (2001b) *Frequently Asked Questions About The National Park Service*. Homepage of the National Park Service, online. Available at: www.nps.gov/pub_aff/e-mail/faqs.htm

NPS (National Park Service) (2001c) *Categories of Natural Signficance and Parks that Illustrate Them*. Homepage of the National Park Service, online. Available at: www.cr.nps.gov/history/catsig/catsig.htm

NPS (National Park Service) (2001d) *American Antiquities Act of 1906*. Homepage of the National Park Service, online. Available at: www.cr.nps.gov/local-law/anti1906.htm

NPS (National Park Service) (2001e) *Historic Sites Act of 1935*. Homepage of the National Park Service, online. Available at: www.cr.nps.gov/local-law/hsact35.htm

New South Wales Parks and Wildlife Service (2000) *Royal National Park*, online. Available at: www.npws.nsw.gov.au/parks/metro/rnphistory.html

Olmstead, F.L. (1865) *Yosemite and the Mariposa Grove: a Preliminary Report*, online. Available at: www.yosemite.ca.us/history/olmsted/report.html

Parker, G. and Ravenscroft, N. (2000) Tourism, 'national parks' and private lands. In: Butler, R.W. and Boyd, S.W. (eds) *Tourism and National Parks: Issues and Implications*. John Wiley & Sons, Chichester, UK, pp. 95–106.

Seibel, G.A. (1995) *Ontario's Niagara Parks. Niagara Falls*. Niagara Parks Commission, Ontario.

USFWS (US Fish and Wildlife Service) (2001) *America's National Wildlife Refuge System*. Homepage of US Fish and Wildlife Service, online. Available at: bluegoose.arw.r9.fws.gov/centennial/index.html

Wade, D.J., Mwsaga, B.C. and Eagles, P.F.J. (2001) A history and market analysis of tourism in Tanzania. *Tourism Management* 21(1), 93–101.

CHAPTER 3
Social Roles of Park-based Tourism

Introduction

Many of the world's national parks and protected areas were originally established to provide people with places for inspiration, recreation and spiritual renewal. As park and protected area systems grew and expanded in scope, protection of biodiversity and cultural heritage also became important rationales for designating areas as parks. In the USA, the initial national parks provided the country with the 'high' culture that many Europeans felt was missing from that nation. The parks shortly became valued in their own right for the 'pleasuring grounds', wildlife, geological features and opportunities to experience, appreciate and learn about wild nature that were scarce in the more developed countries of the late 19th century.

Understanding how to manage parks and protected areas requires not only scientific inventories of the biophysical and cultural attributes contained within them but also a clear realization of their functions and roles in the particular society in which they are situated. The tourism associated with visits to national parks represents an articulation of these roles, a commitment by individual citizens to engage with the parks and experience the values that are protected there. Tourists visit parks and protected areas because there are values in them that they may not experience elsewhere. The roles that parks play in a society and the

Fig. 3.1. Understanding how to manage parks and protected areas requires scientific inventories of the biophysical and cultural attributes, and a clear realization of their functions and roles in the society. The Great Barrier Reef Marine Park of Australia protects the world's largest coral reef complex. It is also critically important to the economy of the state of Queensland, due to the important international tourism economy it stimulates. Heron Island contains both a marine biology research station and an ecotourism resort, thereby fulfilling scientific, cultural and economic functions. Heron Island in Great Barrier Reef Marine Park, Australia.

closely associated tourism that occurs within them serve important functions that are often linked to a national identity or purpose.

Tourism thus serves important social functions, just as parks and protected areas do. Management of tourism in national parks and protected areas is founded on an understanding of what those functions are. In this chapter, we review the various roles and values of parks and park-based tourism in society. We note here that park-based tourism may be only a minor part of the tourism activity in a particular country, and much of that other tourism activity may have little to do with experiencing natural environments or cultural heritage. However, in some countries park-based tourism is a major part of the overall tourism activity. Examples include Tanzania and Kenya.

The Cultural and Social Importance of National Parks and Protected Areas

National parks and other similar protected areas play important roles in society (Box 3.1). The fact that national parks are established usually through a public political process rather than by authoritarian fiat, suggests that the social-political system regards them as important to the

Box 3.1. A family vignette.

The family had high expectations for their visit to Yellowstone National Park, and they anticipated their traditional visit to Old Faithful geyser with cheerfulness. While the father, mother and three children had viewed the explosive eruption of the geyser on many previous visits, the annual trip had become an important part of their family tradition and folklore. The lengthy journey from their home to Yellowstone and the view of the geyser functioned not only to refresh their memories of a magnificent and unique natural feature, but also helped to re-establish and strengthen family bonds. In a sense, their visit to Yellowstone was an American pilgrimage, an experience that was not only recreational and family-building in function, but also provided opportunities for spiritual renewal. It helped to renew forgotten connections to the landscape, and their view of Old Faithful, although in the company of many hundreds of other pilgrims that August afternoon, reinforced their mutual feeling that a higher authority, an overall plan, had provided well for them. While the intricate linkages and processes of this ecosystem and their biological significance were not all that obvious to this typical family, they did understand that without a few thoughtful people acting with courage and foresight over 125 years ago, they would be unable to experience the wonder and beauty of Yellowstone.

The experience of this family is typical of many in Western societies that have become in a very real sense separated from the landscape on which they depend for their livelihoods, at least at an intellectual level. While many parks and protected areas now have additional functions – such as biodiversity protection – the original basis for their designation remains important to society and serves as the basis for considerable tourism activity.

social well-being of the country. Public political processes provide the venues needed to put forth the various arguments, allow for dialogue, deliberation and learning, and establish the political consensus needed to pass enabling legislation and continuing financial appropriations needed for management. During this process of public discourse, the values and importance of a protected area to a society are raised and debated. While not all will agree to the rationales for protecting an area or the values that some may seek through designation, the political process ultimately results in a law and legislative record that documents these values.

In the public discourse that occurs during the debate over establishment, a wide variety of rationales may be put forth for establishment. These range from economic development to the protection of unique geological features or physical processes; from preservation of biodiversity to provision of recreational opportunities. To some, large-sized protected areas represent a 'cauldron of evolution', where ecological processes on an evolutionary scale may continue. Kruger National Park in South Africa, covering about 20,000 km^2, protects an incredible array of wild animal populations that live more or less as they have for millennia. Other areas may protect the cultural heritage important to a nation or even the globe. For example, Pashupatinath, a UNESCO designated World Heritage Site located in Kathmandu, Nepal, is only a few ha in size and protects a sacred shrine of the Hindu religion originally established in the 4th century. This site, one of several located in the Kathmandu Valley, represents the highly developed architectural expression of religious, political and cultural life of early Nepal.

To other people, such protected areas provide significant opportunities for economic advancement as tourists flock to see and experience the values contained within them. Tourist expenditures on routes to the park and in communities adjacent to or within the area may be significant, leading to increased income, alleviation of poverty and opportunities for vertical advancement in tourism businesses. In many cases, the prospect for increased revenues will be a principal political argument advanced for designating an area. Such arguments, however, may not substitute for more fundamental contentions about the inherent values contained within potential protected areas.

Equally significant are the reasons for not protecting an area through formal designation or gazetting processes, for these suggest what other values and needs a society may hold. For example, designation of an area as a national park may eliminate important resource uses, such as grazing or timber production, that a society (or some segments of it) may feel it can ill afford at the time. Such deliberations help to identify the trade-offs that the political system may or may not be willing to make. The dialogue, debate and resolution of the conflict over potential designation intimate the relative worth of differing beliefs and values contained within the proposed protected area. Often, the rationales for protecting landscapes are based on beliefs about the importance of 'nature's guiding hand' in maintaining human health and security. For example, protected areas may often serve as the source of the pure water needed by nearby communities. The protection of watersheds may sustain continuing flows of high-quality water.

The ability of interest groups to get an area designated as a park reflects their capacity to convince the larger social system that the potential values of a park to society outweigh any potential losses, particularly short-term economic losses, associated with its designation. This is exemplified by the establishment of Yellowstone National Park in the USA, when few in the Congress opposed its designation. It was felt that the region was so remote it could not possibly contain resources of any value and thus, there would be no significant economic losses from preservation. At the time, 1872, the Yellowstone region was perceived as being uninhabited. Decision makers became convinced that no minerals or resources of significant value would be found or utilized in such a remote location, and thus agreed to the pleas of a small group of people to protect the region. Of course now the park records about 3 million visits a year, and the surrounding region is occupied by a variety of ranching, mining and timber interests. And the greater Yellowstone ecosystem has also become a popular residential area as people seek regions with high natural amenity in which to live.

The roles of national parks and protected areas vary somewhat by culture and the historical relationship between a culture and the environment in which it is embedded. In some cases, national parks protect sites and features that commemorate events and people that are important in the historical and cultural development of a nation-state. For example, the city of Dubrovnik in the Adriatic country of Croatia was designated as a World Heritage Site in 1979 to recognize the significant Gothic, Renaissance and Baroque churches, monasteries, palaces and fountains contained within the city. Although damaged by armed conflict in the 1990s, the city's historically important features are undergoing restoration to preserve the important architectural elements there.

In other situations, designation of protected areas represents symbolically the importance that particular cultures place on environmental protection. The Great Barrier Reef Marine Park in Australia represents a commitment to protect and preserve the globally significant marine and coral species and the ecosystems and biodiversity represented there. In others, designation represents particular natural processes that exemplify how nature operates. In the USA, designation of Mount St Helens National Volcanic Monument in the State of Washington commemorates the catastrophic volcano explosion in 1980. The Jamestown Flood National Historic Site in the State of Pennsylvania identifies where a small dam broke in the late 19th century after particularly heavy rainfall, flooding the downstream community at a very high cost to human life. Such protected areas may also contain values and opportunities significant not only to the nation in which they are located, but to the world's population as well. In cases where areas contain globally significant features and values, they may be designated as a 'World Heritage Site' such as Glacier National Park in Montana in the USA, recognized because of the outstanding glaciated landscape preserved within it.

Regardless of the exact role and values contained within them, such places often become attractions for tourists seeking a better understanding of the events, people and processes that led to the present situation. Understanding both the roles that protected places play in a specific society and their attraction for tourists (and tourist motivations) is fundamental not only to the management of these places (leading to their protection) but also to the management of those who visit them.

This chapter establishes a working proposition that national parks and similar protected areas are fundamental to the social well-being of a culture. This may be particularly valid for those industrialized societies where the intimate, day-to-day connection with the environment that typified the agrarian cultures from which they arose has changed dramatically. In countries dominated more by an agrarian relationship with the landscape as well as more industrialized ones, the current rationale for protecting areas tends to deal with protecting biological diversity, ecological integrity and preserving habitat for threatened or endangered species before sites are irretrievably committed to other uses. In both settings, providing economic opportunities and enhancing a community's quality of life may also serve as fundamental rationales. Biodiversity values are often viewed as essential to the proper social functioning of a society because they not only provide necessary life-support systems, but also enhance the quality of life of nearby residents and visitors.

National parks and other similar protected areas often preserve places where the meanings that people attach to landscapes deal more with the aesthetic and culturally symbolic than the instrumental or utilitarian. Aesthetic and symbolic attachments in turn deal more with the intangible but important values (such as events, a vision-quest site) than the tangible products (e.g. forage, timber, minerals, petroleum) that landscapes also provide. Such attachments are often ethereal, but widely shared and frequently emotionally arousing. Further, nature- and culture-based tourism represent at the individual scale a manifestation of the larger-scale societal commitment. That is, individuals visit a national park to understand and appreciate the natural and cultural heritage protected by a nation's decision to establish the park.

Cultural Background of Park Travel

While visiting a national park or similarly protected area is an activity in which broad segments of society engage, as a touristic activity it is a relatively new cultural phenomenon. Not only is the idea of nature- and culture-based parks less than 150 years old, the financial and technological means to visit parks for broad segments of society only developed since the end of the Second World War, or about 50 years ago, and only in some regions of the world. While tourism existed prior to the 20th century, it was often limited to the financially elite who travelled primarily in and to Europe for purposes of experiencing high culture or for religious purposes. After the Second World War, with the technological advances of large-scale, inexpensive air travel, more people began visiting a wider array of destinations. As travel activity grew, more people began to visit national parks and protected areas.

Such growth, not only in overall numbers, but also in diversity, typifies tourism in the early 21st century. For example, visits to the continent of Africa are expected to triple over the 20-year period 2000–2020, from 27.8 million to 77.3 million arrivals (World Tourism Organization, 2001). Much of this increase will be to view and experience the outstanding natural heritage located in Africa as well as the unique culture located there.

In some respects, visits to national parks, shrines and sites can be viewed as similar to some religious pilgrimages. Visits to national cemeteries and battlefields reflect the cultural sensitivity to those who fought and died to protect the values that cultures respect. These visits, whether to Old Faithful, a national battlefield site or an historic site can be spiritual, emotionally moving or evenly vividly passionate, evoking feelings of pride, patriotism and wonder.

However, most visits to national parks are motivated by desires to learn about and appreciate natural and cultural history, to gain a sense of history, for the challenges and adventures that nature-dominated environments pose, to seek fun and to use these settings as a place to increase family cohesiveness. Increasing one's physical conditioning may also be important. Protected areas provide opportunities for visitors to develop a sense of perspective, to begin to appreciate that the past played an important role in shaping the present, and to understand that what we now hold dear came because others before us made sacrifices, were worried about the future or were simply far-sighted. Parks are thus highly valued for their opportunities for these experiences.

In the USA, visits to such parks as Gettysburg, Yellowstone, the Arizona Battleship Memorial at Pearl Harbor and the National Capital Parks (Lincoln, Jefferson and Washington Memorials in Washington, DC) may be viewed as similar to a pilgrimage. In these cases, protected areas carry highly charged emotional meanings that are fundamental not only to a society's well-being, but also to understanding the sacrifices that others made for the current generation. Such visits may reinforce patriotic feelings in addition to providing experiential lessons in history. Without such parks, a nation's citizens may lose perspective on the foresight and

Fig. 3.2. Environmental education for school students is a major activity in many protected areas. The exposure to special natural environments during schooling provides students with a lifelong appreciation of nature, culture and protected areas. The desire to learn about and appreciate natural and cultural history is a major background for park travel. Students at the Nature Interpretive Centre, Royal Botanical Gardens, Hamilton, Ontario, Canada. (Photographed by Paul F.J. Eagles.)

sacrifice of their predecessors; they may come to take for granted the freedoms and values they now enjoy.

At least in American society, interest in the cultural heritage of the nation has grown dramatically in the 20th century. As citizens have become more aware of their past, they have pressured Congress to establish more historical parks (as part of the National Park System) and sites, and have increased their visitation to them. Culturally-based tourism is now recognized as a major tourism segment.

The potential pressures that tourism may place on cultural resources are significant, yet such tourism is highly dependent on maintaining the integrity of the site. The International Council on Monuments and Sites (ICOMOS) was established in 1965 as an advocate for protection of culturally important areas. It has increasingly recognized cultural tourism as an important component of its protection and management programmes. At its General Assembly in 1999, it established six principles for encouraging and managing culture-based tourism (Box 3.2).

Basic Cultural Functions of National Parks and Protected Areas

National parks are established because the geographic areas they involve contain socially defined meanings that are significant for at least some sub-populations of a particular society. These meanings may be classified as instrumental, aesthetic and cultural/symbolic (Williams and Patterson, 1999). Instrumental meanings deal with the attainment of some goal, such as production of fibre for wood processing. In the context of this chapter, instrumental meanings include biodiversity values, habitat protection and provision of recreational opportunities.

Box 3.2. Principles of the Cultural Tourism Charter of the International Council on Monuments and Sites (ICOMOS).

Principle 1: Since domestic and international tourism is among the foremost vehicles for cultural exchange, conservation should provide responsible and well managed opportunities for members of the host community and visitors to experience and understand that community's heritage and culture at firsthand.
Principle 2: The relationship between Heritage Places and tourism is dynamic and may involve conflicting values. It should be managed in a sustainable way for present and future generations.
Principle 3: Conservation and Tourism Planning for Heritage Places should ensure that the visitor experience will be worthwhile, satisfying and enjoyable.
Principle 4: Host communities and indigenous peoples should be involved in planning for conservation and tourism.
Principle 5: Tourism and conservation activities should benefit the host community.
Principle 6: Tourism promotion programmes should protect and enhance Natural and Cultural Heritage characteristics.

National parks and protected areas provide important reserves for biological habitats, ecological processes, pure air, clean water and individual species. These functions serve the important role of providing the security that cultures need for maintenance of natural processes important to the survival of human life. National parks and protected areas provide critical habitats for humans to enjoy, appreciate and learn about natural processes.

Finally, national parks and protected areas, through the notion of biosphere reserves, may help society to understand the effects of land uses on the environment and on society itself. Biosphere reserve designations provide for a central, relatively unmodified core of protection surrounded by other lands where modification for economic and subsistence purposes occurs, thus allowing comparative analysis of outcomes of different land management strategies. Often, the central core of a biosphere reserve is a designated national park or protected area.

National parks also provide and protect important recreational opportunities that a society needs to continue to function effectively. These opportunities are needed to relieve the stresses and challenges of a society that is growing in complexity, accelerating in the pace of change, increasing in frustration and increasingly challenged by problems for which there are no easy answers. Recreation provides roles and leads to benefits important not only to the individual, but to society as well, including cognitive development, enhancement of self-esteem, increased social cohesiveness and so on. These values are fundamental to constructive functioning of society, and national parks and protected areas can contribute in positive ways by providing the settings where these benefits occur.

Many of the original national parks in North America, such as Yosemite, Mt Rainier and Rocky Mountain in the USA, and Banff and Jasper in Canada were designated in part to protect their scenic and aesthetic values. During the latter half of the 19th century, American and European artists had become increasingly fascinated with the scale and magnificence of the western North American landscape. Through such romanticist-Hudson River School artists as Albert Bierstadt, Thomas Cole and Thomas Moran, the splendour and sublime character of this landscape was represented in dozens of paintings and sketches. In English Canada, the work of Tom Thompson and the Group of Seven helped to break the domination of European images in art and establish a uniquely Canadian-nationalistic view of the Canadian wilderness. Americans and Canadians became fascinated with the aesthetic values of this landscape and, through a series of congressional and parliamentary actions, acted to preserve it. This focus on aesthetics led to additional designations, but has also resulted in management challenges as park organizations attempted to maintain landscapes at particular points in time, when, in fact, landscapes are dynamic entities.

The variety of roles can be illustrated by a few examples. In the UK, some parks, such as Brecon Beacons National Park in Wales, protect agrarian landscapes and attempt to preserve the landscape in its current condition. Sagarmatha National Park in Nepal preserves opportunities for adventure and challenge for both trekkers and mountaineers. The Statue of Liberty National Historic Site in the USA preserves part of the country's national identity. Great Barrier Reef Marine Park in Australia protects important biodiversity values. Glen Canyon National Recreation Area in the USA provides opportunities for water-oriented recreation.

In addition, parks may represent a symbolic step for particular cultures that revere life, even if a species has no known value as a resource, amenity or for scientific purposes. There is often a strong sense that all species should be protected and provided with the needed habitat, even if the probability of finding a genetic or biological value in the future is relatively small.

Tourism's Role in Society

Like parks and protected areas, tourism plays several crucial roles in contemporary society. Of course, the basic purpose of tourism is to provide individuals with opportunities for escape, stress-release, challenge and adventure, strengthening family cohesiveness, learning about and appreciating one's natural and cultural heritage, and experiencing new cultures. These particular benefits of tourism are fundamental to healthy individuals and societies.

However, tourism often plays other important roles in various societies. The dominant role, as perceived by many, is tourism's potential for the economic and financial opportunities it provides. National parks and protected areas offer opportunities for economic development in two ways: (i) by supplying the resources and attractions that non-residents will visit, and thus lead to an increase in economic opportunity for local residents; and (ii) by providing an amenity-rich backdrop to communities that businesses will find attractive when making relocation decisions. Such economic benefits accrue to both the individuals that directly participate and the community, as the community's economy becomes more vibrant as a result of tourism spending. (Both economic functions are discussed in greater detail later in this book.)

However, there are other roles that tourism plays which are often overshadowed by its obvious economic role. These roles and benefits accumulate at individual, household, community and national levels. Obviously, as noted earlier, participation in tourism allows people to see and experience cultures, both their own and others; it enhances communication between people of different cultures, providing opportunities to increase awareness and understanding of different traditions, religions and rituals. At the community level, tourism provides opportunities for towns and villages to demonstrate these rituals, helping in some cases to preserve them for the future. Communities may also engage in

Fig. 3.3. Artists capture the moods of a park. Their art often becomes part of the cultural identity of a park and its community. Protected areas play important roles for society, with art being a major driver of public cultural appreciation. Artist in the Monteverde Cloud Forest Reserve, Costa Rica. (Photographed by Paul F.J. Eagles.)

partnership activities with villages located in another nation as a result of tourism activity. At the national level, tourism, because of the opportunities for dialogue and understanding it permits, may lead to increased opportunities for peace and stability among nations. This is the hope of the International Institute for Peace through Tourism, an international non-governmental organization. Its mission is to foster and facilitate 'tourism initiatives which contribute to international understanding and cooperation, an improved quality of environment, the preservation of heritage, and through these initiatives, help to bring about a peaceful and sustainable world' (IIPT, 2001). The relationship between tourism and peace is two-directional, as noted by Egyptian President Hosni Mubarak:

> Tourism and peace are intertwined. The former cannot flourish without the latter. Moreover, tourism fosters understanding and peace, which support one another and enable continuity. It also broadens opportunities for cultural exchange and encompasses nations from all parts of the world.
>
> (Mubarak, 2001)

Another role of tourism, increasingly, is to assist in protecting the resources on which it is based through revenue to park management agencies, such as the Saba Marine Park in the Netherlands Antilles or the Community Baboon Sanctuary in Belize, as discussed in other chapters. The notion of ecotourism (Box 3.3) has as part of its founding philosophy the idea that the visitation that occurs in a protected area does so partly to protect the area. This can be through increased understanding of the biodiversity and cultural values protected there, which, in turn, leads to activism in protecting those values as well as helping to finance the management of the area.

Box 3.3. Ecotourism: a new role of park-based tourism.

The development of concerns about the impact of tourism on both the natural environment and local communities and cultures has led to new forms of tourism, under the rubric of ecotourism. Ecotourism involves activities and experiences that explicitly attempt to minimize impacts, benefit local communities and protected areas, and encourage appreciative and learning activities on the part of tourists. This form of tourism has been particularly attractive in developing countries that have learned from the mistakes of the more developed ones.

The development of ecotourism thus represents a new value or role of tourism for society, one where the benefits lie both at the individual level and at the community level. This evolution of the meaning of tourism has also encouraged greater sensitivity to environmental and cultural impacts by the mainstream tourism industry.

Case Study Number 3: Wakatobi Marine National Park (Indonesia) and Project Wallacea (UK)

Ecotourists, Researchers and a Non-governmental Organization Assisting with Park Creation and Management

Wakatobi Marine National Park contains 1.39 Mha of marine, coastal and tropical forest environments in the Wallacea region of Sulawesi between Borneo and New Guinea. The area contains very important biodiversity. The park was declared in 1996 and is the second largest marine national park in Indonesia. The park has 55 rangers involved in the protection of the park and in the implementation of a management plan. The park rangers have dramatically reduced the level of destructive and illegal fishing techniques.

Operation Wallacea is a British-based organization with an associated charitable trust to which it pledges a proportion of any surpluses created. Operation Wallacea operates a dive and marine research centre in the national park, and began biological surveys of the Wallacea region in 1995. On recognition of its ecological importance, the group lobbied for the creation of a national park. This lobbying, working with a counterpart in Jakarta, the Wallacea Development Institute, stimulated the Indonesian government to create the national marine park in 1996.

Each year Operation Wallacea brings substantial numbers of scientific volunteers to the national park. Each volunteer pays for the experience. In 2000, there were 300 university students, together with 20 scientists, three professional photographers, two artists, expert trackers and forest support teams, diving staff and logistics team members working on a wide range of projects. The volunteers stay an average of 5 weeks in the area, between June and October. Each volunteer is responsible for a research project dealing with some aspect of marine biology, ecotourism research, forest ecology, and wildlife management or community conservation. In 2001, students and professors from the UK and the USA worked on 50 different research projects. In addition, volunteer naturalists, divers and photographers completed a range of research and community development projects. The project is designed so that the visitors have positive economic impacts on the local communities. Approximately 60 local families gain all or a significant proportion of their income through employment, contract work or provision of supplies. Overall, 50% of all monies paid by the volunteers is spent in the local communities. Five local people are supported for each volunteer visitor.

The project constructed an environmental education centre operated by the project, which provides reef biology courses to over 1000 children each year. Funding provides for community work in order to provide sustainable economic avenues for the local people.

Operation Wallacea is one of the largest examples of a coordinated, volunteer-based, park research project in the world. The park managers report that the scientific research findings and the presence of the scientists in the park help enormously in achieving the park's management objectives.

Operation Wallacea successfully manages to bring the benefits of ecotourism to a grass-roots level. This empowers local people by providing a new and lucrative source of income that did not previously exist. It has enabled local communities to see the value of protecting natural resources, such as rainforests and coral reefs, rather than depending on their exploitation.

Wakatobi National Marine Park and Project Wallacea was granted the British Airways/ World Conservation Union award for park tourism in 2000. It is a superb example of park visitors, scientists, researchers, NGOs and park managers working together for environmental conservation and community development (websites: www.operationwallacea.win-uk.net/mainmenu.htm; www.opwall.com; www.wakatobi.com/backgrd/research.htm).

Changing Cultural Conventions

Of course, societies and cultures are not static. Human society experienced significant changes over the latter half of the late 20th century and as we enter the 21st century, there is no indication that the pace, scale and complexity of change will itself slow. A number of these changes are significant for management of national parks and tourism, including globalization of economies, growth in international travel and visits to parks of all kinds, population and demographic changes, human migration at both the international level and within nations, ageing of the population in industrialized countries, and evolving attitudes towards the environment. These changes imply that parks and protected areas established by one government will be valued by others living elsewhere. For example, many of those visiting Sagarmatha National Park in Nepal are Europeans and Americans, just as many of the visitors to Jasper National Park in Canada reside in Japan.

Two major concepts that have developed since the mid-1970s as a result of these trends are important here. One concept is that of sustainability, which as it developed was a melding of economic, environmental and quality-of-life goals. The concept was given life by the World Commission on Environment and Development in its final report *Our Common Future* published in 1987. The Commission identified several major concerns confronting the world's population – population growth, food security, species and ecosystems, energy, pollution from industry, and urban growth – and suggested that initiating sustainable development strategies would help to alleviate poverty, reduce environmental degradation and protect the quality of life. The report generated considerable discussion worldwide in numerous venues and among many disciplines, including tourism and protected area management.

Sustainable development was defined as 'development that meets the needs of the present without compromising the ability of future generations to meet their own needs'. While the World Commission on Environment and Development identified and defined sustainable development, a large number of questions remain, such as how does an agency, institution or government protect its environment, alleviate poverty, enhance economic opportunity and improve the quality of life for its citizens while maintaining options for the future? What is the role of tourism in this effort? How do national parks and protected areas fit in? How should they be managed sustainably?

Such questions have encouraged academics, tourism planners and interested members of the public to seek tourism that is sustainable. Sustainable tourism is generally defined as touristic activity that is small in scale, protects the integrity of local cultures, minimizes negative impacts on the environment and, yet, benefits local economic conditions. The evolution of the concept of sustainable tourism has focused efforts on developing planning and management strategies that attempt to meet these requirements.

The second concept is the idea that actions to manage protected areas must be acceptable to the dominant social group using or interested in the area. The notion of attending to the social acceptability of management actions has developed out of a growing recognition that management and planning processes must be more inclusive of the people affected by decisions if they are to be successful in changing the future. Protected area planning and tourism development over the course of the 20th century had become dominated by a scientific, expert-driven model that tended to exclude the local community and informal (experiential and emotional) forms of knowledge. Many countries adopted a North American model of protected area designation, which resulted in removal of indigenous populations from within the protected area boundaries. Both approaches led to a situation of increasing alienation of those for whom protected areas were designed to serve and increased resistance to protected area decisions, particularly within adjacent communities.

To gain acceptability, management must not only be more inclusive of who participates in the decision making process, but also recognize that forms of knowledge other than technical/scientific knowledge are useful. In a sense, management decisions must be endorsed by those directly affected by them to be viewed as legitimate and to be supported in order for them to be implemented. Such support and endorsement come about only with the involvement of the affected public in planning and management processes. Being concerned about the social acceptability of management actions does not mean that management is conducted by popularity votes or that the public makes administrative decisions; rather that the social and cultural basis for protected areas is recognized, those affected by decisions are involved in their making and public concerns are directly incorporated into decision making. This may be difficult for some park planners and managers to accept, because it means that decision-making power, once the domain of the bureaucratic elite, is now diffused among the public.

Changing Park Functions Over Time

North American parks provide a good example of how the social functions and values of national parks have changed over the long period of formalized park designations. These changes are partially mirrored in other countries. The first American national park was established in 1872 with Yellowstone National Park. This date falls in the midst of the Industrial Revolution in Western society. The Yellowstone park legislation was modelled after the earlier Yosemite grant (to the state of California) in 1864, which ceded the Yosemite Valley to the state for 'public use, resort and recreation', thus establishing parks as recreation areas. The colony of New South Wales in Australia in 1879, Canada and the province of Ontario in Canada, both in 1885, followed the USA lead in designating their first national parks.

The National Park System in the USA was initially composed of a few 'gems' ideally representing the most magnificent natural and unmodified features in the country. Importantly, congressional legislation establishing these areas, while mentioning preservation, also described parks as 'pleasuring grounds'. Early interpretations of national park functions tended to reinforce this concept and differed significantly from today's concepts. Runte (1997) argued that the primary rationale for national parks in the 19th century was to put American culture on an equivalent footing with that of the Europeans: the magnificent natural features were substitutes for the castles, cathedrals and works of art as national symbols or monuments often found in Europe. The initial American national parks also provided both a nation-building claim and reasons why Americans should care about the areas west of the Mississippi River.

Nash (1973) contended that the concept of wilderness (as we now define it) and the preservation of ecological processes were not motivating factors in early national park establishment. Instead, preservation was viewed as an outdoor museum concept, with nature perceived as static. Different components of nature were frequently defined as 'good' or 'bad' and dealt with as such. For example, large predators were systematically exterminated from some park settings (as late as the 1950s) in the early 20th century simply because, as predators, they were viewed as 'bad' since they killed the 'good' animals like deer.

During the first half of the 20th century, as industrialization and the federal government exerted more influence over American life, the function and role of the park system was considerably expanded. A host of different types of areas – such as the historical (or 'cannonball' parks preserving American Civil War sites), national recreation areas, national seashores and national wildlife refuges with purposes that were not as constrained as the original parks – were added to the system. Many of these areas were located nearer urban areas and provided greater opportunities for outdoor recreation. These areas did not necessarily preserve magnificent, pristine landscapes, but rather were established to provide opportunities for a diversity of outdoor recreation and educational experiences. The national lakeshores, seashores and recreation areas thus permitted more facilities and development while historic sites and battlefields emphasized event and site integrity. This burgeoning diversity suggested a broadening of social understanding of the roles that protected areas serve in contemporary American society.

Parks are very much the products of the culture that creates them: they are social institutions in the truest sense of the word. Their purpose and management policy reflect the dominant values and needs of the society in which they are emplaced. Thus, as American society changed, so did the function of parks. In the 1950s and 1960s, the dominant need seemed to be for entertainment, and visitors were presented with an array of choices during their trip to a park. These included: firefalls (Yosemite National Park), ski areas (Rocky Mountain, Banff, Jasper), bleachers (seating) for watching bears claw their way through garbage (Yellowstone), marinas (Glen Canyon National Recreation Area), lightshows (Roosevelt National Recreation Area) and bars (nearly every national park).

But cultures change and evolve over time, values shift, priorities change. What was once appropriate and acceptable may no longer be so. What once was not acceptable, may now be so. Changes in roles and functions, as they evolve over time, may not be particularly acceptable to specific interest groups; these changes may be the source of contentious debate over park values. Predicting the future and what might be acceptable is difficult because the change occurring around the world in the late 20th and early 21st centuries is systemic in character. The underlying assumptions about park and protected area planning to meet future needs must be examined in the light of the fundamental value shifts now occurring.

National parks originated at the height of the Industrial Revolution. But we are now an information-based society, the characteristics of which result in four major implications for parks and protected areas. First, the national parks in the industrial era were promoted and managed under the implicit assumption that the more people that visited the areas, the greater the flow of public benefits. Thus, an early measure of successful management of parks and protected areas was the rise in visitation. This assumption is now severely questioned. As Driver and others note (1991), success may be defined more by what happens to a visitor (in terms of increased learning, expectations met, etc.) than by how many people visit a park. Management now focuses on the kind or quality of experience visitors receive (and who receives these benefits) during an engagement at a park.

Second, the parks now host new types of recreational engagements. While these types of experiences initially held a high adventure or risk recreation element (skydiving, hang-gliding, river rafting, rock climbing, for example), they have more recently evolved into experiences that have greater emphasis on learning and interaction with local populations, through an ecotourism focus. Ecotourism programmes provide visitors with opportunities to interact and learn about local cultures and customs, in addition to the natural environment, while contributing to their protection.

A third area in which national parks are assuming a new role concerns redefinition of the term ‘preservation’. The industrial era definition was oriented around the concept of parks as outdoor museums. Emphasis was often placed on maintaining vignettes of park landscapes as they existed with designation. However, parks are no longer established and managed to preserve a static landscape (unless the park has more of a historical purpose), but as areas in which the operation of natural ecological processes is preserved, thus recognizing the dynamic character of nature. This distinction is significant. No longer are there ‘good’ or ‘bad’ components of nature; death, predation, natural disturbances (such as fire, flood, avalanche) and naturally occurring population changes are viewed in a disinterested context in the sense that parks may be best preserved if nature can take its own course as much as possible. Visitor developments once viewed as desirable because they helped to increase visitation are now viewed as disruptive, intrusive and inappropriate (e.g. the infrastructure developments at Fishing Bridge in Yellowstone intrude on grizzly bear habitat and were removed).

A fourth new role that parks have now assumed concerns their value for science. The Laguna Blanca Reserve in Argentina is an example. Originally a small national park, the reserve now totals 973,000 ha and includes a sizeable lagoon as well as mountain and highland vegetation communities. The reserve contains a core area of 163,000 ha that remain unmodified, a buffer zone of 48,000 ha to shield the core from human impacts and a transition zone of 762,000 ha where human activities occur. As noted by the Man and the Biosphere Programme of UNESCO (2001), such reserves are designed to address three functions: (i) a conservation function, to contribute to the conservation of landscapes, ecosystems, species and genetic variation; (ii) a development function, to foster economic and human development that is socio-culturally and ecologically sustainable; and (iii) a logistic function, to provide support for research, monitoring, education and information exchange related to local, national and global issues of conservation and development.

Thus, national parks as scientific reserves not only provide opportunities to understand the nature contained with them – including ecosystem structure and function and population dynamics – but also may serve as genetic reserves where discoveries of species or DNA helpful to humans may occur. In addition, such reserves also provide opportunities for examining the effects of human activities on the environment. Many of the areas serving as internationally recognized biosphere reserves were originally designated as national parks. The original designation remains, with the biosphere label overlaid.

What this example shows is that the function of national parks, protected areas and the tourism associated with these areas changes and evolves as the values of the larger society in which they are embedded also change; subcultures appear, evolve and disappear at different rates. These changes also suggest continuing conflict over purpose and management, thus making the job of a manager, or tourism promoter, challenging.

Conclusion

Parks and protected areas serve a variety of functions and roles in society. The specific benefits that societies value are highly dependent on each culture. Major values include biodiversity protection, landscape aesthetics preservation, education and learning, challenge and adventure, and opportunities to enhance important social values such as family cohesiveness. Tourism represents one behavioural articulation of these values but, again, the importance and distribution of these values,

Fig. 3.4. Public instruction in the ecological sciences is a major focus of the interpretive programmes of many parks. One goal of these programmes is to produce an ecologically literate populace. The specific benefits that societies seek are highly dependent on each culture. US culture values both wildlife and education, both of which are represented in the environmental education centre in San Francisco Bay National Wildlife Refuge. Environmental Education Centre in San Francisco Bay National Wildlife Refuge, USA. (Photographed by Paul F.J. Eagles.)

functions and roles are different from one society to the next and may change over time.

A full understanding of these values is needed to properly manage parks. In a sense, they represent broad goals and philosophies; if they are not understood, how can a park or protected area be managed efficiently and effectively to meet these values? Often, however, there is considerable discretion on the part of managers as to how these values will be protected and managed. While the goal may be understood, there may be very different alternatives to consider. Societies, through political processes, establish goals and values; managers through administrative processes, determine how those goals and values will be achieved.

Sustaining the values venerated by a society has become increasingly important as citizens, scientists and managers come to more fully understand the impact of human activity on the environment and as their appreciation for the cultural and natural heritage protected by parks grows. At the same time, growing interest in management by the public implies that managers must be more inclusive of the public in planning processes, a topic we turn to in the next several chapters.

References

Driver, B.L., Brown, P.J. and Peterson, G.L. (1991) *Benefits of Leisure*. Venture Publishing, State College, Pennsylvania.

IIPT (2001) *International Association for Peace through Tourism*. Homepage of IIPT, online. Available at: www.iipt.org/main.html

Mubarak, H. (2001) *President Mubarak's Concept of Tourism*. Homepage of The Ministry of Tourism of Egypt. Available at: www.touregypt.net/mubarak.htm

Nash, R. (1973) *Wilderness and the American Mind*. Yale University Press, New Haven, Connecticut.

Runte, A. (1997) *National Parks: the American Experience*. University of Nebraska Press, Lincoln, Nebraska.

UNESCO (2001) *What is a Biosphere Reserve?* Homepage of the MAB Program, online. Available at: www.unesco.org/mab/brfaq.htm

Williams, D.R. and Patterson, M.E. (1999) Environmental psychology: mapping landscape meanings for ecosystem management. In: Cordell, H.K. and Bergstrom, J.C. (eds) *Integrating Social Sciences with Ecosystem Management: Human Dimensions in Assessment, Policy and Management*. Sagamore Publishing, Champaign, Illinois, pp. 141–160.

World Tourism Organization (2001) *Tourism 2020 Vision – Africa*. World Tourism Organization, Madrid.

Further Reading

Barzetti, V. (ed.) (1993) *Parks and Progress: Protected Areas and Economic Development in Latin America and the Caribbean*. IUCN, Gland, Switzerland.

Chambers, E. (ed.) (1997) *Tourism and Culture: an Applied Perspective*. State University of New York Press, Albany, New York.

Conlin, M.V. and Baum, T. (eds) (1995) *Island Tourism, Management Principles and Practice*. John Wiley & Sons, Chichester, UK.

Hall, C.M. and Lew, A.A. (eds) (1998) *Sustainable Tourism: a Geographical Perspective*. Addison Wesley Longman, Harlow, UK.

Harrison, L.C. and Husbands, W. (eds) (1996) *Practicing Responsible Tourism: International Case Studies in Tourism Planning, Policy, and Development*. John Wiley & Sons, New York.

Honey, M. (1999) *Ecotourism and Sustainable Development: Who Owns Paradise?* Island Press, Washington, DC.

Innskeep, E. (1991) *Tourism Planning: an Integrated and Sustainable Development Approach*. Routledge, London.

Lindberg, K., Wood, M.E. and Engeldrum, D. (eds) (1998) *Ecotourism: a Guide for Planners and Managers*, Vol. 2. The Ecotourism Society, North Bennington, Vermont.

Wearing, S. and Neil, J. (1999) *Ecotourism: Impacts, Potentials, and Possibilities*. Butterworth-Heinemann, Oxford.

Chapter 4
Planning for Tourism in National Parks and Protected Areas: Principles and Concepts

Introduction

In earlier chapters we have suggested that tourism is a significant and essential component of contemporary human life, argued that national parks and other similar protected areas play important but varying roles in culture, and that park tourism should not just happen, but should be explicitly managed, directed and controlled. These arguments can only lead to the conclusion that planning is central not only in providing adequate protection for the principal product base (the local cultural and environmental heritage) of tourism but also to the business components of the tourism industry (e.g. lodging, food, services, guiding and outfitting) itself.

We note here that tourism is often portrayed as potentially carrying the seeds of its own demise. Indeed, unmanaged tourism development, promotion and marketing can lead to severe and irreversible social and environmental consequences: tourism is often accused of loving a park to death. High levels of unmanaged tourism visitation may result in significant disruptions of wildlife movements, destruction of wildlife habitat, impacts on water quality and so on. Elizabeth Kemf details how tourism development on the Greek Island of Zakynthos threatened the nesting grounds of the loggerhead sea turtle (Kemf, 1993). Such development can destroy the very foundation of a young and sensitively developed nature-based tourism industry. Large-scale tourism developments are often criticized for their negative effects on water and vegetation, large requirements for energy and heavy production of solid and liquid wastes. The secondary and tertiary social and environmental effects of tourism development are often difficult to discern and may occur offsite or take a long time to display themselves. If not properly managed, these negative consequences can potentially threaten the quality of the environment within which tourism facilities are located. Thus, by negatively affecting the environmental or cultural features that tourists find attractive, the industry may unexpectedly damage its very product base.

While park tourism may very well lead to a number of negative consequences, benefits may result as well, as we have noted in other chapters. Yet, the benefits of park tourism may not be optimized, or negative impacts minimized, without good planning, planning that considers the consequences of action or, in some cases, the lack of action on the agency's ability to meet a desired future condition.

Of course, facilities may enhance social and environmental qualities as well, through appropriate planning, design and mitigation actions. Indeed, one could argue that tourism development should not occur unless these qualities are improved. While the criticism of tourism as leading to negative impacts is often well supported, it is frequently the ineffective management of tourism that allows such impacts to occur.

Many parks and protected areas were originally designated without much consideration for their impacts on local communities or indigenous peoples or without much thought to natural boundaries. Indeed, even Yellowstone National Park was thought to be an 'empty' place when it was established in 1872, yet evidence uncovered later indicated that it was used as a hunting area and spiritual place for local Native North American Indian peoples. Parks and protected areas do indeed affect local communities and indigenous populations in both positive and negative ways. Good park tourism planning acknowledges these potential effects, accounts for them and acts on them in decision-making processes.

Moreover, tourism frequently has been used as a political strategy to gain public support for national parks and protected areas. In the USA, the Northern Pacific railroad was largely responsible for designation of the Yellowstone region as a park. In Canada, the push of the Canadian Pacific Railway into the Rocky Mountains precipitated the creation of Banff National Park. In many other American and Canadian parks, railroads were encouraged to develop tourist facilities to allow more people to visit. Stephen T. Mather, the first Director of the US National Park Service and James Harkin, first Commissioner of the Canadian Dominion Parks Branch, actively pursued a policy of tourism promotion and development, primarily as a political strategy to gain public support and, therefore, needed government appropriations to fund park management.

More recently, tourism has been viewed as a means for gaining the revenue and building the 'ownership' within nearby communities needed to protect critical habitats and species. For example, the black howler monkey in Central America is threatened; a primary cause of the monkey population decline is loss of habitat. In the country of Belize, the Community Baboon Sanctuary was designated, involving several communities and consisting of private lands to protect habitat. Fees earned from tourists visiting the sanctuary are returned to participating communities and farmers as an incentive for habitat protection (CBS, 2000). The issues of tourism, protected areas and local communities are addressed in more detail in Chapter 8.

While such strategies still may work effectively, there will always remain a tension between tourism development and protection of important cultural and natural heritage values. These tensions raise fundamental questions about the contexts and purposes of park planning. To address these questions in this chapter, we first provide a more thorough analysis of challenges of 'park planning'. We argue that the park planning challenge is not simply one of protecting parks from development and unmanaged use, but is much more inclusive, considering the communities adjacent to or within it and the tourism industry to which it is linked.

In this discussion, we recognize that park planning is an inherently 'messy' task, thus requiring new and more effective approaches to planning. Park planning is founded on some general principles and processes concerning not only the content of the park plan, but also how the plan itself is developed. We conclude the chapter with a discussion of the more inclusive issue of involving communities in the planning process by articulating some general objectives of what planning should strive to accomplish.

What Park Planning is All About

The issue of planning for parks and protected areas has grown in complexity over the years. It has done so because our knowledge of the intricate relationships between humans and their environment has accelerated rapidly, increasing our awareness of the consequences of human actions. This complexity has also been influenced by the expanding diversity of groups with interests in parks and their broadening range of views towards park management. We have also realized that past

Fig. 4.1. The management plan is the overall policy statement for a park. This meeting allows staff from head office, the park and foreign aid groups to discuss the implications of all aspects of the proposed park plan. The issue of planning for parks and protected areas has grown in complexity over the years, thereby requiring the involvement of many people from a diversity of backgrounds. Management plan meeting at Ruaha National Park, Tanzania. (Photographed by Paul F.J. Eagles.)

approaches to protected area planning, based in particular disciplines using an expert-driven model of planning no longer result in the types of knowledge adequate for the decisions that must be made. But this knowledge is not limited simply to the effects of human activity on the special values that protected areas preserve, but also includes the web of interactions between people and their environment at a variety of temporal and spatial scales. In this section, we review the character of the park planning (its inherent 'messiness'), the various substantive dimensions it involves and the role of a management plan in dealing with the problem.

Park planning is a 'messy' process

If nothing else, the issue of sustainability, which has been widely accepted as a precept guiding natural and cultural resource management, forces us to consider planning as a way of maintaining human and other forms of life on this planet. Sustainability, in the sense of intergenerational equity and maintenance of the natural capital that may be required by future generations, represents a redistribution of economic and political power to those generations. Achieving sustainability requires thinking in longer timeframes and potentially forgoing present income and opportunities for those that may occur in the future. It means, in the words of the World Commission on Environment and Development (WCED, 1987), which stimulated much of the present discussion on sustainability, providing for the needs of the present while maintaining options for future generations to meet their own needs. This is a tall order, and one that cannot easily be met. It involves considering a bewildering array of social preferences and values (both for present and future generations), institutional structures and barriers, philosophical outlooks, forms of knowledge and conflicting perceptions of what is important.

The notion of sustainability is similar to the mission of many national parks: to protect an area for the enjoyment of future generations. But national parks play other roles important to future generations as we noted in Chapter 3. They provide reserves to protect biological diversity. They supply needed remnant habitats for threatened species. They serve as areas for families to gain cohesiveness. They function as irreplaceable outdoor learning laboratories. They may protect important grazing and plant collection areas for people residing in the local area. They may contain important historical, cultural and spiritual sites. Understanding all these missions is fundamental to any planning process, yet the variety and complexity of these missions reveals the dynamics of underlying social values and suggests that not only are park problems often complex, but they will require a wide variety of expertise and forms of knowledge for their resolution.

This protection does not occur simply through a gazetting process. Drawing lines on maps is only a symbolic gesture if it is not accompanied by a process to retain, sustain or restore the values for which the park was established. In some cases, those values exist in abundant quantities and are of high quality. In other cases, the full public benefit of those values can only occur after restoration of the ecosystem or the cultural environment. To a very real extent, planning involves choosing a future and designing pathways to it. In both cases, understanding what futures are possible and which are desirable are the required first steps in active protection. Taking the requisite action needed to achieve desirable futures is a further step in their protection.

In either case, such values can only be maintained in the face of constant population and exploitation pressures if a socially acceptable planning, management and enforcement framework is present.

Planning helps to determine which of many possible futures is a desirable one. Figure 4.2 graphically illustrates this aspect of planning. It shows that the future is very much a function of the pathways, management actions and choices we make today. It suggests that there are many possible futures, but some of those may be unacceptable or undesirable. From the range of potential futures, we select one that appears to be desirable. Once this desirable future is selected, we implement policy to ensure that it actually transpires. Thus, at its most fundamental level, planning is needed to ensure that the desired future becomes a reality. Thus, a park *plan* documents the chosen desired future, and the general policies selected to arrive there (see Fig. 4.3).

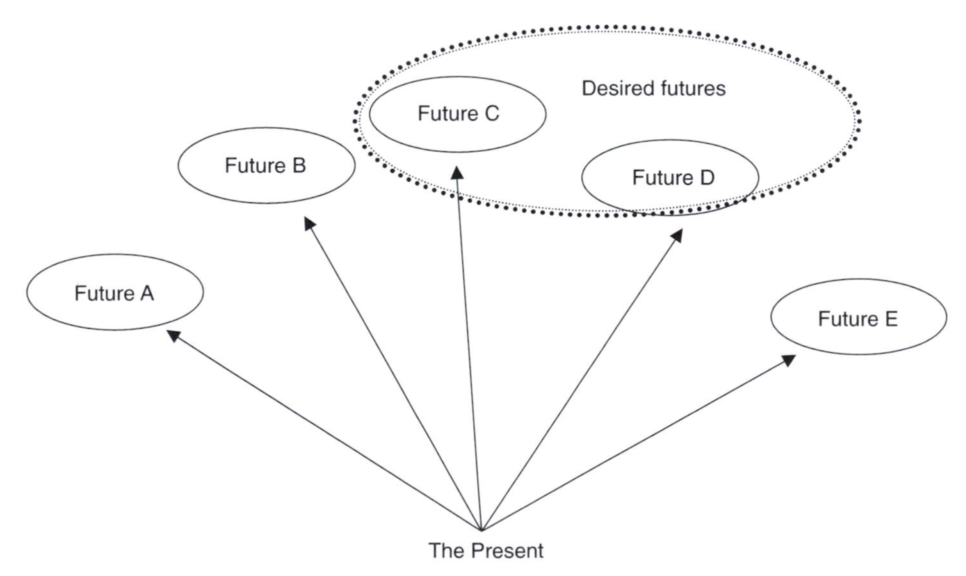

Fig. 4.2. The future is very much a function of the present and the actions we take. This figure shows, schematically, that a particular park may have several potential future conditions, only two of which are desirable. Planning is a process that selects which future, among the many possible, is desirable and the appropriate pathway to it. The arrows suggest that different pathways lead to different futures.

Fig. 4.3. Planning is needed to ensure that the desired future becomes a reality. Most cities in the English world planned and developed natural parks close to the city centre. The model for these parks was the park system of central London. Toronto Island Park, Toronto, Ontario, Canada. (Photographed by Paul F.J. Eagles.)

Management identifies the means – the policies, actions and pathways – of attaining those futures and maintaining them. Planning often involves identifying the range of appropriate management actions, actions that are predicted to be effective and efficient. Management determines what actions occur when, by whom and at what cost. Management also secures the resources needed to monitor implementation, evaluate results and adapt actions as needed. A *park management plan* details the methods, actions, timetables and resource allocation (people, money and facilities) needed to attain and maintain the desired future. The management plan represents a strategy to implement the decisions made by park managers and stakeholders about what potential future is desired.

Enforcement ensures that the actions determined to attain or maintain futures are implemented. We often think of enforcement as applying to individuals breaking laws and regulations. But enforcement also applies to management in the sense that planning processes identify outcomes that can be used to hold agencies accountable for the actions they agree to take. Therefore, enforcement must occur both at the institutional level and at the individual level.

Finally, as we note in Chapter 5, *monitoring* is conducted to ensure that our visions of the future are being achieved through management actions. Monitoring involves systematic and periodic measurement of key variables that indicate progress in achieving management goals. Figure 4.4 schematically represents the relationships between these important components.

Planning

is a process of analysis and assessment that leads to a

Management Plan

which is a formalized document stating goals, means to attain them (actions) and monitoring strategies.

Management plans are

Implemented

through yearly commitments of funds, people and policy changes.

Implementation is

Monitored

to ensure that the goals established in the planning process are achieved.

Such monitoring not only ensures timely progess but is also a form of

Enforcement

to ensure not only that visitors follow rules, but that park organizations live up to their commitments.

Fig. 4.4. The relationships between planning, management, implementation, monitoring and enforcement. All are key elements of protecting the natural and cultural heritage for which park organizations have a stewardship responsibility.

Unfortunately, there is often little consensus on which future is the desirable one. Society is highly fragmented into a variety of values and perspectives on what roles parks should assume and what objectives should be developed for any given park. Various stakeholders and interest groups may hold differing preferences for park goals and values. While most national parks have legislatively mandated objectives, the interpretation of these objectives has shifted over the last few decades in response to changing social preferences. There may be different interpretations of mandated goals. Competing interest groups often contest stated objectives for individual parks. Frequently, enabling legislation or decrees appear intrinsically conflicting (e.g. preserve while providing enjoyment). The interests of local communities and national communities may diverge. Failure to agree on and accept specific objectives is a major reason why plans fail to be implemented.

The role of science and technical knowledge is limited in situations where goals are contested because the fundamental issue confronting the park is one that centres on different values rather than the 'how to's needed to solve a particular problem. Thus, the issue of restoring predator populations in some American parks and protected areas, such as timber wolves in Yellowstone, dealt more with the conflicting ideologies of the role of parks in society than the actual science of restoration. While science can inform planners of the consequences of alternatives in this type of conflict, its role is limited because the issue is one of disagreement over goals, a conflict that is properly within the domain of public policy.

Consideration of information needs at longer timeframes and larger spatial scales typical of ecosystem-based management regimes have also led to a situation where scientific disagreement over causes and effects has risen to new levels. At the same time, we increasingly understand that there is a lot that we do not understand. Tracing the effects of tourism development and park management practices (particularly the second and third order consequences) is difficult, time consuming and expensive. Many effects may not be observable for a long period of time and many others occur outside the boundaries of a park.

An example is the fire exclusion and suppression policies often practised in the 20th century by North American park managers. Since the early 1920s, park managers attempted to exclude both naturally occurring and human-caused fires in national parks and other similar areas. They were successful in this attempt, for a while. These policies, however, led to significant changes in vegetation patterns, patch sizes and fire intensity. Areas that had formerly experienced high frequency, low intensity fires became subject to catastrophic fire as vegetation grew, changed and fuel accumulated. These effects were not generally observable until after 50–75 years of fire exclusion policy when, in 1988, a series of large, catastrophic fires raged through Yellowstone National Park and other places in the northern Rocky Mountains of the USA. The fires were much larger and more intense because of a combination of an accumulation of fuel over the decades and large scale meteorological events (e.g. *El Niño*). The fire suppression policy initially resulted in fewer fires, but more recently led to much more intense burns. Foresters and managers employed by park and land management agencies were often surprised by the magnitude and intensity of these fires. They eventually concluded that fires need to be managed, which means that they are permitted to burn under some conditions, suppressed in others and in certain cases, initiated to manage fuels.

In any given planning situation, there may be disagreement among scientists or uncertainty concerning relationships between causes and effects. Relationships between causes and effects underlie actions proposed in the planning process. If there are disputes or uncertainty about them, it may be difficult to garner public support for their use. Uncertainty can be reduced by treating management as an experiment so that more information is gained. Unfortunately, the large spatial scales and long timeframes often involved in park management often prohibit scientifically defensible experimental designs.

Situations with varying amounts of agreement on goals and scientific uncertainty are depicted in Fig. 4.5. This figure shows, simplistically, four situations in which park managers may find themselves. Situations that are characterized by agreement on goals and agreement on cause–effect relationships may be termed tame problems. Tame problems occur when there is a consensus about what future is desired and we know what actions will achieve that future. Planning processes for tame problems are well developed and form the backbone of how planners normally tackle policy issues.

Situations where goals are not contested but cause–effect relationships are disputed may be termed mysteries. These mysteries require additional emphasis on science because it is unclear whether a particular management action will lead to an expected outcome. Here, science would play a very important role in providing planners with the information needed to select among specific alternative actions and policies.

Where there is disagreement about goals but scientists agree on cause–effect relationships, wicked problems result. Such problems generally represent the value conflicts that we spoke of earlier. Here, planning processes would emphasize identifying the values and ideologies in conflict and then negotiating resolution of the conflict through some type of accommodation of interests. Science is not involved, except perhaps in identifying and displaying the values in conflict.

Finally, messy situations occur when there are disagreements about both goals and cause–effect relationships. In this situation, there is a high level of uncertainty in the planning situation; it is likely to be fluid, dynamic and highly contentious. Making progress on a plan will require a high degree of both facilitation skills and leadership in encouraging research to help reduce uncertainty.

The planning processes for each of these problem situations vary as the key words in Fig. 4.6 show. What is important here is that the planning process chosen is appropriate for the situation. Traditional approaches to planning based on science and expert opinion are termed rational-comprehensive planning; they are appropriate only for tame problems where there is a single goal on which a consensus exists.

		Social agreement on goals	
		Agree	**Disagree**
Scientific agreement on cause–effect relationships	**Agree**	Tame	Wicked
	Disagree	Mystery	Messy

Fig. 4.5. Park planners are confronted with a variety of planning situations as described by the two contextualizing variables of social agreement on goals and scientific agreement on cause–effect relationships. The trend is towards planning in messy situations (adapted from Thompson and Tuden, 1987).

		Social agreement on goals	
		Agree	**Disagree**
Scientific agreement on cause–effect relationships	**Agree**	Rational-comprehensive	Negotiation
	Disagree	Adaptive	Learning, consensus building

Fig. 4.6. Appropriate planning emphasis for different situations.

Such planning processes are not appropriate for wicked or messy situations. The character of the planning challenge is such that it is not so much a problem of information as of values; as such, technical information may do little in designing an appropriate pathway to a desired future. Park planners more and more frequently find themselves embroiled in messy situations: settings where goals are contested, sub-problems are linked, issues are contentious, and solutions are only temporary because the context is continually evolving. In such messy situations, planning processes emphasize dialogue, mutual learning and consensus building over scientific expertise, technical information and expert opinion.

The Interconnected Character of Park Planning

Like other issues, planning at the interface of national parks and tourism is largely dependent on how issues and challenges are defined or framed. The questions we ask and how we ask them influence the responses and plans that are ultimately developed. Planning for protected areas is not solely for preserving the values for which a national park was initially created. Protected areas are not geographically isolated islands immune to outside influences and frozen in time. More likely, protected areas are islands of environmental, cultural and scenic quality located in a sea of rampant development and change. Obviously, development and land use practices outside the park influence what happens inside the boundaries; ecological and social processes do not necessarily follow administrative boundaries. Ecological processes and conditions occurring within the park also affect what happens outside. Therefore, park planning must be regarded as having strong elements of regional planning processes.

If we assume, as we should, that tourism and community development are inextricably tied to parks, then other issues, such as establishing ownership by the local community in the planning process, increasing the quality of life of local peoples, enhancing economic opportunity, developing capacity for accommodating tourists and providing economic incentives for enhancing local interest in the park, are directly involved in park planning. Casting the issue of protected area planning – particularly in places that have local communities or indigenous populations residing within the park or immediately adjacent to it – as developing actions only within park boundaries seriously miscasts and excessively reduces the subject matter of protected area planning.

The Issue of Capacity

While national parks are established to protect culturally significant environmental and heritage values, the problem of planning for them is often depicted, in a tourism context, as 'how many visitors are too many?' This question implies a need to establish visitor carrying capacities for parks and tourism destinations. Originally, the term carrying capacity, in a visitor management context, was defined as the number of visitors that can be accommodated at a park without degradation of the biophysical quality of the area. This definition has evolved over time, including such meanings as how much change (defined in terms of human impacts) from natural conditions, but still carries the implication of determining how many visitors are too many.

This way of defining the problem frequently leads to such statements as parks being 'loved to death', 'overused' or 'overcrowded'. This particular way of framing the problem has led many managers to limit tourist numbers, only to determine later that such limits come at very high costs to visitor choices, freedom and experiences, with benefits to biophysical conditions that are questionable (McCool and Lime, 2001). While the consequences of tourism use of national parks are somewhat related to use levels, more influential are factors such as visitor behaviour, type of tourism development, season of use, management approaches and biophysical characteristics. All are significant factors in influencing the intensity, duration and type of impact experienced.

The issue of tourist numbers and their resulting impacts is also influenced by the management structure and philosophy of the park organization. For example, for much of their history, North American parks were financed entirely through appropriated tax revenues that had accrued to the national governments. In this case, the park organization was required to submit and adhere to a fixed budget, one heavily influenced by prior budgets. Typically, visitor numbers were a minor element in the budget allocations even though many park services are directly related to visitation levels. Visitors were then often perceived as a drain on limited funding and not as a source of revenue. Such a situation puts managers into a difficult position. Budgets were often insufficient to handle existing visitor loads, and year-to-year increases in visitor use incurred even more challenges.

Management structures in other areas provide a different view of visitor numbers. Throughout eastern and southern Africa, national parks

Fig. 4.7. The issue of tourist numbers and their impacts is influenced by the management structure and philosophy of the park organization. The goal of developing an economically viable management structure is assisted by the airstrip within Kruger National Park. Kruger is a major international ecotourism destination, thereby contributing positively to the park's income and the local community's economic development. Skukuza Airport, Kruger National Park, South Africa. (Photographed by Paul F.J. Eagles.)

and game reserves do not receive appropriated money from national governments' general funds. All park funding is earned from park entrance fees, special charges on accommodation, fees for specialized programmes and facilities, and donations. Therefore, understanding visitor use levels is critical to the fiscal health of the agency and level of protection of the resource. This perspective leads to an attitude that visitor use is a benefit, albeit to the revenue for the park organization.

In summary, there is a variety of influences on how the issues of park visitation and carrying capacity are perceived. Often, these influences reflect wider societal philosophies on how government services will be funded and who should pay for them.

A park planning approach then would be to redefine the issue of park capacity for visitors to one of determining what biophysical and social conditions are acceptable or desirable. Such a definition focuses planning and management effort on the outputs of management, recognizes that many park planning decisions are value laden (what is desirable/acceptable is a function of values and beliefs) and implicitly suggests that a variety of interests and values are represented in planning processes so that common definitions and agreements on desirable/acceptable can be developed. We will turn more explicitly to this question in Chapter 5.

Process Principles for Park and Protected Area Planning

Park planners are confronted with two major challenges when initiating planning for a specific area. First, they must design a planning process that is scientifically and technically sound, inclusive of those stakeholders affected by or involved in the park, conducted in a timely and efficient manner and results in specific, acceptable management actions that will achieve the goals for the area. Second, the outcomes of the planning process, i.e. the actions identified, themselves must be placed on a firm foundation of understanding of not only the values of parks but also their role in a particular society. In this section and the next, we present several principles guiding development of planning processes to achieve both these goals. We first turn to a set of process principles.

Principle 1: There must be explicit criteria for evaluation of alternatives

Park and protected area plans are designed to find routes to a desired future. For any given route or pathway, there are usually several alternatives available as schematically represented in Fig. 4.2. Thus, planners must select a set of alternatives that, in their judgement, is the 'best' pathway to the desired future. The criteria used to select the 'best' pathway are fundamental to the planning effort. What criteria are chosen will determine which alternatives are selected as the preferred. Criteria may include equity, efficiency and effectiveness concerns. They may reflect desires to improve the quality of recreational offerings. Or they may deal with amount of environmental or social impact. The planning staff may develop criteria, and ideas for such criteria may be collected from the public involved in the planning process. Once criteria are selected, the alternatives are evaluated and a decision rule is determined. For example, a decision rule might state that the alternative with the least administrative cost to implement would be selected.

Principle 2: Park planning requires a variety of forms of knowledge

There are many different forms of knowledge available to planners and decision makers. These include scientific/technical, managerial experience, emotional, anecdotal or individual experience and community forms. No one form of knowledge is intrinsically better than any other form. All forms are legitimate. Some sources of knowledge that will be used in planning are accurate and reliable, other sources may not be. Since action in society requires a variety of actors (planners, administrators, politicians, engineers, local community leaders, etc.) and each prefers certain forms of knowledge over others, each form must be acknowledged, accepted and acted on appropriately.

Principle 3: Park planning in messy situations requires public participation

Participation of the public is a hallmark of contemporary protected area planning. Implementation of plans is highly dependent on public appreciation and support; neither is developed when agencies write plans hidden from public purview. Historically, planning was viewed as a technical process about which the public had no expertise and all that was required was for the public to respond to the alternatives developed by the planners. However, that view of planning and public participation is disappearing rapidly.

The public holds a variety of roles in park and protected area planning, such as a source of innovative ideas, a testing ground for potential management actions and validating technical information. In addition, since public funds are normally expended to finance management of parks, there must be a feeling of responsibility or ownership in the planning process in order for the affected public to develop the consensus needed for appropriating monies to implement the plan. The question is therefore not *if* the public will be involved, but *how*. This involves understanding the type of power held by the various groups affected and identifying the specific roles that the public will assume in the planning process. Some interest groups, for example, may hold 'veto' power over implementation of plans because of their political activism.

Planners, when developing the planning process, need to develop a public involvement plan that specifies the objectives of public involvement, how the public will be involved, when, and how public input and comments will be treated. Public participation objectives might include identifying the social acceptability of various alternatives, creating ownership or responsibility for the park and its plan, ensuring that the public included is representative of diverse interests, creating opportunities for learning (involving dialogue among planners, scientists and the public) and enhancing relationships among planning process participants. These questions are particularly important when the planner is confronted with a contentious planning context.

Principle 4: Uncertainty and risk must be acknowledged in the park plan

Our knowledge about the consequences of managerial actions is limited; often such actions lead to unintended consequences – effects that were not or could not be foreseen. Uncertainty is the condition when probabilities cannot be assigned to effects. Risk occurs when we know the distribution of effects and can assign a probability to their occurrence. For example, campsite closures are often not completely successful; some people may camp at a site even if it is posted as closed. Such conditions, where consequences are not known or may be known but only with a level of probability, are known as decision making under conditions of uncertainty and risk, respectively. The planning process should detail how uncertainty and risk will be treated. For example, monitoring of compliance with a campsite closure will inform managers of the effectiveness of the posting. Different ways of informing or persuading the public to comply with the closure could then be tested. Monitoring soil conditions, such as soil compaction, will help to determine if the closure was effective.

Principle 5: The plan must address temporal and spatial scale mismatches

Scale mismatches occur when one ecological or social process operates at one scale and another process occurs at another. Ecological processes such as vegetation succession may operate at a decadal level, while financial appropriations to manage vegetation occur at a biennial level. Since budget priorities change, there is no assurance that, over a decadal period, the priorities will provide the funding needed at a level adequate for vegetation management. Thus, plans must be flexible, adaptive and responsive to changing institutional environments. The plan should contain mechanisms to deal with such potential changes.

Principle 6: Park plans include statements of the personnel and financial requirements for implementation

A fundamental purpose of planning is to link knowledge to action (Friedmann, 1987). Plans developed without implementation strategies are bound to fail simply because they do not articulate the timing, responsibility, costs and trade-offs required to achieve the identified goals. Allocation of personnel and financial resources will influence which alternative is selected as the preferred one. Thus, each management plan must contain a statement describing the budgetary and personnel needs for implementation.

Foundational Principles of Park Planning in Messy Situations

There is also a set of content principles important for planners in messy situations.

Principle 7: The natural and cultural heritage forms the basis for all other values and benefits associated with a protected area

This principle suggests that protection of the heritage within the park takes primacy in decisions. Yet, it does not imply that there is necessarily and always a zero tolerance for any development or activity. Providing recreational opportunities and associated development means that trade-offs will and must occur. Development does not happen without some type of impact on the values protected in a park. Planners and managers are thus confronted with the question of how much impact will be desirable or acceptable.

Principle 8: Recreational activities in the protected area are dependent on the maintenance of desired conditions, yet provide substantial monetary and societal benefits to participants, local community and park management

This principle carries two important implications. First, recreation activities provided should only be those dependent on or directly linked to the natural and cultural features protected in the park. For example, high quality scuba-diving and snorkelling in a marine park is highly dependent on a biophysical setting that has a minimum of impacts present. Secondly, provision of recreation opportunities may result in substantial financial and economic impact, in both positive and negative ways, on the local community and the park management organization. Understanding the consequences of proposed management actions suggests that these effects must be included in environmental assessments.

Principle 9: The protected area management organization exists to protect the natural and cultural heritage through active management of recreationists and provision of learning opportunities for the local community

Parks and protected areas are important not only for those values they preserve, but also as recreation destinations for visitors and as a source of learning opportunities for those visitors and the local community. This suggests that parks are directly linked to community well-being. In this respect, parks serve several functions. First, they often provide the scenic backdrop important as an element of quality of life often sought by individuals. Second, parks and protected areas may be the source of clean air and pure water important to the health of individuals living within the community. Third, parks serve as resources for educational systems where students may learn about the cultural and natural heritage protected therein, and how to provide the stewardship for parks. Finally, parks, because of their attraction for non-residents may be the source of substantial economic benefits. Other values and functions of parks are described in Chapter 3. The specific functions, objectives and benefits of an individual protected area are often described in organic legislation or government policy for the area. Typically, they are further detailed in the planning process.

Principle 10: Protected areas exist within a larger social and environmental context that requires active community involvement and understanding

As we noted earlier, parks are not isolated entities, but are linked to their larger social and natural environment in a number of ways. Parks and their management cannot effectively protect the values for which they were established without active community involvement in their management. Community involvement brings appreciation, trust and a sense of ownership and pride in the park, qualities that are important in securing agreement on necessary management actions and building the political constituency needed for public funding.

Summary of Principles

Management plans thus serve several functions as they incorporate the above principles. First, they identify a desired future and a path that, under current and projected situations, will lead to that future. If nothing else, plans serve to help change the future towards a more desirable one. Secondly, plans institutionalize the desired future and the apparent most effective path to it. By so doing, there is less implicit subjective bias in management's decision-making processes. Thirdly, the plan makes explicit the preferences and value systems of those involved in the planning process. Ideally, such processes have incorporated and addressed the concerns and interests of the various park stakeholders. By making plans and planning processes explicit, biases are revealed and decisions are more likely to be defendable. In a sense then, plans are a contract between the public and the bureaucracy that manages the park. This leads to greater accountability on the part of the management organization to fulfil the direction given in the plan and implement the policies and actions stated in it. This also allows the public as a political constituency to lobby for the necessary funding allocations to implement the plan.

Visitor Management within the Park and Protected Area Plan

One aspect of park planning that is receiving increased attention is the idea of integration. In this context, integration means that all aspects of managing a park are considered in an initial planning document. This is necessary because different aspects of park problems and threats are highly linked. For example, managing visitors in many parks has implications for management of wildlife. In the Amboseli National Park of Kenya, for example, a major lodge was located in prime wildlife habitat, thus exacerbating conflicts between people and wildlife. Locating major facilities cannot be conducted in isolation of the management of natural processes. Yet, planners must be careful not to overwhelm the planning

Fig. 4.8. The interaction of park visitors and wildlife is an important element in most park plans. Such interaction involves both desirable and undesirable aspects. Park visitors like to observe large mammals. However, when a bear enters a campsite and raids the camper's food supply, management intervention may be necessary. Black bear in campsite in Killarney Provincial Park, Ontario, Canada. (Photographed by Paul F.J. Eagles.)

process with too much complexity or detail so that the process 'bogs' down and cannot be completed. The document may be termed a general management plan, policy plan or something similar.

Such integration, however, does not mean that the wildlife, fisheries, vegetation, fire and visitor management chapters are written separately and then stapled together. This is not integration. Integration means that the various specialists, appropriate scientists and members of the public jointly define the issues to be addressed in the plan, with an understanding of the contribution each discipline and form of knowledge can contribute to resolving the issue. And while there will be separate chapters on each of the relevant topics, those chapters may be jointly authored and include references to the linkages with other planning document sections.

Within this context, visitor management will probably be one of the issues pursued in a park planning process. It is often a complex issue itself and, thus, is often dealt with separately from other components. Yet, one cannot address management of visitors in most parks without also addressing how they interface with many other issues such as fire or wildlife (Fig. 4.8). One cannot determine how grizzly bears, for example, will be managed without considering the effects visitors have on them.

Visitor management often focuses on two major questions: (i) what can be done to enhance the quality of a visitor's experience? and (ii) how can the impacts of visitors be managed to acceptable levels and for desirable outcomes? The first question needs to be addressed from a park tourism perspective: visitors expect high-quality recreational engagements. This does not necessarily mean fancy facilities, expensive programmes, or porters and servants wherever one goes. It does mean that planners and managers are aware of visitor expectations and, where appropriate and consistent with the protected area goals, attempt to meet them. The second question must be addressed because many parks are established to protect or preserve important cultural or natural values; visitors entering such places may have a negative impact on those values, so management is needed to reduce those impacts. Management is also necessary to enhance the desirable impacts of visitors, such as community support, cultural appreciation and revenue generation.

Management techniques can be grouped by the degree to which they intrude on a visitor's experience (such as use limits or law enforcement) and the degree to which they are obvious or subtle. A general philosophy in North American parks has been a preference for techniques that are not intrusive and are more subtle, putting more responsibility on the visitor. Thus, there is an emphasis on visitor education and information rather than regulation and restriction. In Chapter 5 we will examine these questions in depth through a description of the Limits of Acceptable Change planning process.

Case Study Number 4: St Lawrence Islands National Park (Canada)

Park Zoning Using a Standardized Zoning System

Parks Canada Zoning Background

The national parks zoning system in Canada (Parks Canada, 1994a) is an integrated approach by which land and water areas are classified according to ecosystem and cultural resource protection requirements (Table 4.1). Their suitability and capability for providing opportunities for visitor experiences is also a fundamental component of developing a park zoning strategy. Zoning is one part of an array of management strategies used by Parks Canada to assist in maintaining ecological and cultural integrity by providing a framework for the area-specific application of policy directions, such as for resource management, appropriate activities and research. As such, zoning provides direction for the activities of park managers and park visitors alike. The application of zoning requires a sound information base related to the function and sensitivity of ecosystem structure, as well as the opportunities and impacts of existing and potential visitor experiences.

St Lawrence Islands National Park

St. Lawrence National Park is located on the St Lawrence River and consists of a series of islands scattered along an 80-km stretch (Parks Canada, 1994b). Each island is managed as a distinct environment with varying facilities ranging from docks, trails, camping facilities, a boat launch, interpretive displays and day-use areas. This particular park is a good example of zoning because it manages four of the five zoning classes. There are nine special preservation areas that the park manages in an effort to protect areas that represent both natural and cultural heritage resources of the Thousand Islands region. Class 2 is the wilderness class, and it is not an appropriate class for St Lawrence Islands because the park only encompasses 869 ha. It is, in fact, Canada's smallest national park. A wilderness zone requires a landmass of 2000 ha or more. There are many areas that have been zoned as Class 3, and these areas provide a variety of opportunities for visitors to experience the park's natural values through low-density outdoor activities, with development of appropriate facilities and services. These facilities include picnic shelters, primitive campsites, trails, interpretive panels, toilets and docks. The outdoor recreation areas (Class 4) are capable of accommodating outdoor recreation opportunities and related facilities. A full range of visitor uses is permitted in this zone. The least protective zone for a park is Zone 5, park service areas. These areas provide a place for visitors' services, support facilities and the administrative functions required to manage and operate the park.

Table 4.1. Zoning system summary for Parks Canada.

			Management framework	
Zone class	Zone purpose	Boundary criteria	Resources	Public opportunity
1 Special preservation	Specific areas or features that deserve special preservation because they contain or support unique, rare or endangered features or the best examples of features	The natural extent and buffer requirements of designated features	Strict resource preservation	Usually no internal access. Only strictly controlled and non-motorized access
2 Wilderness	Extensive areas that are good representations of each of the natural history themes of the park and which will be maintained in a wilderness state	The natural extent and buffer requirements of natural history themes and environments in areas 2000 ha and greater	Oriented to preservation of natural environment setting	Internal access by non-motorized means. Dispersed activities providing experiences consistent with resource preservation. Primitive camping areas. Primitive, roofed accommodation including emergency shelters
3 Natural environment	Areas that are maintained as natural environments and which can be sustained with a minimum of low-density outdoor activities and a minimum of related facilities	The extent of natural environments providing outdoor opportunities and required buffer areas	Oriented to preservation of natural environment setting	Internal access by non-motorized and limited motorized means, including in the north, authorized air charter access to rivers/ lakes, usually dispersed activities, with more concentrated activities associated with limited motorized access. Rustic, small-scale, permanent, fixed-roof accommodation for visitor use and operational use. Camping facilities are to be at the semi-primitive level

Continued

Table 4.1. *Continued.*

Zone class	Zone purpose	Boundary criteria	Management framework	
			Resources	Public opportunity
4 Recreation	Limited areas that can accommodate a broad range of education, outdoor recreation opportunities and related facilities in ways that respect the natural landscape and are safe and convenient	The extent of outdoor opportunities and facilities and their area of immediate impact	Oriented to minimizing impact of activities and facilities on the natural landscape	Outdoor opportunities in natural landscapes or supported by facility development and landscape alteration. Camping facilities will be to the basic serviced category. Small and decentralized accommodation facilities
5 Park services	Towns and visitor centres in certain existing national parks that contain a concentration of visitor services and support facilities as well as park administration functions	The extent of services and facilities and their area of immediate impact	Oriented to emphasizing the national park setting and values in the location, design, and operation of visitor support services and park administration functions	Internal access by non-motorized and motorized means. Centralized visitor support services and park administration activities. Facility-based opportunities. Major camping areas adjacent to, or within, a town or visitor centre to the basic serviced category. Town or visitor centre

Planning for Local Communities and Tourism

We noted earlier that planning for national parks and tourism occurs within a dynamic context, but is linked to tourism and community issues. Park planning must be cognizant of the relationship between national parks and the communities that are located immediately adjacent to them, and sometimes within them. Such communities typically are heavily dependent on the park for their economic base. Many residents may use the park as an important recreation area and park policies themselves are influenced directly by the political interests of local communities. At some time, national policies may diverge from the interests of the local community.

Nowhere is this as problematic as in Banff National Park in Canada and Yellowstone National Park in the USA. In Banff, a local community is located within the park and has developed as an upscale tourist destination attracting hundreds of thousands of international visitors each year. The community is replete with ski areas, lodging, restaurants, shopping malls and evening entertainment sites. The impacts of the community on the park are significant, politicized and complex. Likewise, park management policies have an impact on the community and its burgeoning tourism industry. For example, the Banff town site is located in prime wildlife habitat. As the town's population, park visitation and associated highway traffic increase, impacts on wildlife populations also increase and become ever more inconsistent with the park's mandate. At some point, such impacts will no longer be acceptable as Parks Canada struggles to meet its mission of preserving the natural heritage located in the park.

West Yellowstone, Montana, a major entry into Yellowstone National Park, is a relatively small community that over the last 25 years has diversified its narrowly-based summer season tourism industry to encompass a strong winter season based on snowmobiling and snowmobile rentals. While there are numerous snowmobile opportunities located *outside* the park, much of the industry is based on snowmobiling *in* the park. In both cases, tourism promotion strategies, developments and actions within the community influence the expectations and behaviours of visitors entering the parks.

Such snowmobiling activity has become increasingly controversial and managers grapple to deal with interactions with park wildlife, particularly bison. Bison populations rose dramatically over the last few decades and the animals increasingly leave the park during the winter for feed. Their ability to move outside the park is influenced by the presence of the compacted snow used by snowmobilers to access the park. Some of the bison also carry the disease brucellosis, which, under certain circumstances, may be transmitted to domestic cattle, making them unfit for human consumption. Thus, park management policies aimed at reducing snowmobiling activity to influence bison movements may directly affect the community by changing its economic base. Since such policies are normally developed at a national level or, at least, place national interests above local ones, such local communities develop a sense of alienation and lose trust in national governments.

A third example is Sagarmatha National Park in Nepal. Sagarmatha is home to Mt Everest, and about 20,000 people take the trek to the base camp annually. The trek follows traditional pathways that connected the villages located within the park. How the park is managed directly affects these communities, and their economic, cultural and political interests influence how the park is itself managed. When the park administration adopts management actions that are counter to traditional lifestyles, it finds enforcement difficult. Restrictions on gathering wood for cooking and warming fires in homes and lodges is an example. As tourism grows, demand for fuelwood increases, but the only sources are located in the park. Thus, restrictions on fuelwood gathering affect local communities and the tourism industry.

Community participation is one proposal to overcome such problems. In community participation, the community, through its leaders, directly engages the park authority in planning processes and assists in accommodating the various interests and values at stake. In this approach, local and experiential knowledge is treated as legitimate as scientific knowledge during the search for acceptable actions. Often, the problem of planning in national parks has been posed as balancing local community and national interests. But this assumes – through the use of the term balance – that these interests are not only different, they are also competing. Through community participation, one learns that the interests are often widely shared – for example, communities really do not want the resource base on which their economy is founded to be degraded – so the question is not one of balance but one of accommodation and integration.

Conclusion

Planning for tourism in parks and protected areas cannot be conducted in isolation of local communities located in or adjacent to those areas. The linkages between parks and communities are simply too strong to ignore in such planning. Protected area planning involves two aspects: the planning process itself and the content of the plan. Both must be founded on sound scientific, technical and social principles and concepts. In this chapter, we have proposed that the primary task of planning is to identify a desired future, obtain agreement on it, and develop a management regime to ensure that this future is achieved.

Such plans, of course, must account for the institutional capacity of park organizations for implementation. Grandiose plans requiring large amounts of funding, significant increases in personnel and many new facilities, roads and trails may be impossible to implement. The financial structures, management effectiveness and enforcement capability of the park organization strongly affect the success of planning.

These futures involve both communities and the protected areas adjacent to them. Thus, while park planning is often viewed as a technical planning process involving only the area within the park, it is actually a social/political process that impacts, negatively and positively, on nearby communities. Given this, it is imperative that those affected by plans are closely involved in their development.

We did not suggest any particular planning process to follow. Such processes are normally defined by each protected area management agency. Planning issues and contexts vary; no one process will be applicable to every issue and situation. We will describe a visitor management planning process in Chapter 5 that can be adapted and modified for specific situations.

Box 4.1. What is a community?

In Chapters 4 and 8, we discuss linkages between parks and communities. However, there are many definitions of what a community is, as rural sociologists can testify. The traditional definition of community is a place – such as a town, village or city – where a group of people resides. It used to be that the people living there often shared many values. While this is still true to some extent, there is often a diversity of perspectives towards park management and other social issues in any such spatially defined community.

Another way of defining a community concerns the sharing of values regardless of where a person may live. Thus, there are communities of interest, such as membership in a preservation organization. The people in these communities may rarely have face-to-face interaction, but through the mass media and the Internet do share values.

Finally, the term community may more generally refer to people living within a region or other places in a nation, regardless of what values they may or may not share.

In this book, we use the term community to encompass each of these meanings, simply because a lot of people are interested in how tourism in parks and protected areas is managed, regardless of where they live or the groups for which they hold membership. From the context of the discussion, the type of community to which we are referring should be clear.

References

CBS (2000) *Community Baboon Sanctuary Belize.* Homepage of Community Baboon Sanctuary, online. Available at: www.ecocomm.org/cbs.htm

Friedmann, J. (1987) *Planning in the Public Domain: From Knowledge to Action.* Princeton University Press, Princeton, New Jersey.

Kemf, E. (1993) Tourism versus turtles. In: Kemf, E. (ed.) *The Law of the Mother: Protecting Indigenous Peoples in Protected Areas.* Sierra Club Books, San Francisco, California, pp. 186–193.

McCool, S.F. and Lime, D.W. (2001) Tourism carrying capacity: tempting fantasy or useful reality? *Journal of Sustainable Tourism* 9, 372–388.

Parks Canada (1994a) *Guiding Principles and Operational Policies.* Hull, PQ, Canada. Parks Canada, National Parks Directorate.

Parks Canada (1994b) *St Lawrence Islands National Park Management Plan.* Hull, PQ, Canada. Parks Canada, National Parks Directorate.

WCED (World Commission on Environment and Development) (1987) *Our Common Future.* Oxford University Press, Oxford.

Further Reading

Anderson, D., Lime, D.W. and Wang, T.L. (1998) *Maintaining the Quality of Park Resources and Visitor Experiences: a Handbook for Managers.* Cooperative Study Unit, University of Minnesota, St Paul, Minnesota.

Bacow, L.S. and Wheeler, M. (1984) *Environmental Dispute Resolution.* Plenum Press, New York.

Clark, R.N. and Stankey, G. (1979) *The Recreation Opportunity Spectrum: a Framework for Planning, Management and Research.* USDA Forest Service, Pacific Northwest Forest and Range Experiment Station, Portland, Oregon.

Kuss, F.R. (1990) *Visitor Impact Management, a Review of Research.* National Parks and Conservation Association, Washington, DC.

Lime, D.W. (ed.) (1996) Congestion and Crowding in the National Park System. Minnesota Agricultural Experiment Station, University of Minnesota, St Paul, Minnesota.

Manidis Roberts Consultants (1997) *Developing a Tourism Optimization Management Model (TOMM), a Model to Monitor and Manage Tourism on Kangaroo Island, South Australia.* Manidis Roberts Consultants, Surry Hills, New South Wales.

Manning, R.E. (1999) *Studies in Outdoor Recreation: Search and Research for Satisfaction.* Oregon State University Press, Corvallis, Oregon.

National Park Service (1997) *VERP: a Summary of the Visitor Experience and Resource Protection (VERP) Framework.* NPS, Denver, Colorado.

Susskind, L. and Cruikshank, J. (1987) *Breaking the Impasse: Consensual Approach to Resolving Public Disputes.* Basic Books, New York.

Susskind, L., McKearnan, S. and Thomas-Larmer, J. (eds) (1999) *The Consensus Building Handbook: a Comprehensive Guide to Reaching Agreement.* Sage Publications, Thousand Oaks, California.

Thompson, J.D. and Tuden, A. (1987) Strategies, structures and processes of organizational decision. In: Thompson, J.D., Hammond, P.B., Hawkes, R.W., Junker, B.H. and Tuden, A. (eds) *Comparative Studies in Administration.* Garland Publishing, New York, pp. 197–216.

CHAPTER 5
Management of Visitors in National Parks and Protected Areas

Introduction

We noted in Chapter 4 that designating an area and drawing a boundary around it cannot adequately protect national parks and protected areas if planning and management do not accompany the gazetting process. Management of the factors that threaten the values for which the area was established is needed. In addition, most parks and protected areas need management to enhance values, such as when tourism has become an integral component of the park. Since most parks and protected areas were established to accommodate some type of recreational use and promote visitor learning, these parks require a visitor management strategy to ensure that the opportunity to achieve these values is optimized and that such uses do not lead to unacceptable levels of negative impact. Management is required to enhance such values as learning, appreciation, adventure and challenge, as well as ensuring acknowledgement of community and tourism interests. As we noted earlier, managing visitors and tourists in parks is but one component of a large and related set of policies that must be established and implemented to protect the qualities and values for which a park has been designated.

Management of tourists within the boundaries of national parks and protected areas has important implications for communities and landscapes outside those boundaries. Management of snowmobiles in Yellowstone National Park in the USA, for example, influences the economic viability of nearby communities such as West Yellowstone and Cooke City, Montana, which in the 1980s and 1990s expanded their winter-use tourism season by catering to snowmobiling occurring both within and outside the Park. Changes in national park policy limiting snowmobile use inside the Park have significant economic effects on those communities and potentially place additional pressures on landscapes outside the Park. Thus, management of visitors inside the Park boundary must account for and acknowledge effects occurring or induced in communities outside the boundary.

In this chapter, we examine management of visitors inside the park boundary by reviewing systems approaches specifically designed to enhance visitor experiences and control negative impacts. In Chapter 6, we follow this presentation with a discussion of the specific management techniques that can be used to influence and control visitor behaviour; primarily information and education, allocation and rationing systems, and measures dealing with resolution of conflicts between types of visitors.

We begin our discussion with brief statements of several principles that are fundamental to managing visitors. Following this, we examine the ideas of diversity of recreation opportunities and zoning as a management tool, two notions that are important to any park management regime. We next turn to a description of the popular Limits of Acceptable Change system of managing visitor impacts and opportunities. The emphasis in this chapter is on understanding principles, concepts and rationales for visitor management systems so that specific techniques and approaches may be adapted to local situations.

Fundamental Visitor Management Principles

Visitor management is guided by several principles that are fundamental to sound park management. These principles are as important to planning as information about the biophysical aspects of a protected area.

Understanding park visitor characteristics, motivations and expectations is key to effective management policies

There are three uses of such visitor information that enhance the capability of park management organizations. First, information about visitors, their characteristics, motivations and expectations is needed to develop programmes and provide opportunities that enhance their recreational experiences. Without an understanding of what experiences and opportunities visitors seek, managers cannot manage the setting, within legislative and policy constraints, to increase the opportunity to achieve those experiences. This is particularly important in situations where the local community and park organization desire to promote tourism in the park. Second, such information is needed to effectively communicate to visitors information about the park, its objectives, mandates and mission. This communication is also a foundation of management actions aimed at changing visitor behaviour. Third, such information is helpful to managers in understanding the social/political acceptability of proposed management actions. Fundamental to visitor motivations in visiting national parks is a sense of freedom, where the locus of control appears to be within the individual. When visitors perceive regulations as unnecessarily intrusive or interfering with their motivations they are likely to oppose them, and compliance will not be complete.

Visits by tourists to national parks occur during their leisure time. An essential ingredient of leisure time is the freedom to choose where, when and how to recreate. Thus, the control over one's behaviour lies internally rather than externally as in work environments where supervisors and co-workers determine what will be done, when and how. This need for freedom is important in designing management actions to minimize obvious effects on this pursuit. While freedom of control is an important value underpinning visits, this does not mean that there should be no regulation, only that managers may want to try less intrusive mechanisms first.

Visitor-related developments generally represent both the best opportunity for appreciation of the park and the key internal threats to its biophysical or cultural integrity

Visitor-related infrastructure in parks (roads, trails, interpretive centres and signs, restrooms, campgrounds, car parks, etc.) have three primary purposes. First, such developments facilitate recreational uses by protecting park values and limiting the negative impacts of those uses. For example, wooden walkways across a meadow limit the soil compaction, erosion and vegetation loss that would result from traditional unsurfaced pathways. Second, facilities enhance recreational experiences, appreciation and learning opportunities by providing the physical infrastructure necessary for them. For example, a signed trail interpreting natural history may be viewed as a facility necessary to enhance opportunities for learning and appreciation. Third, park facilities may be designed to provide for the safety of park visitors. Well-designed facilities, such as barriers, keep visitors away from such dangers as cliffs, fast water, hot springs and wild animals (Fig. 5.1). However, such facilities represent threats to the natural or cultural heritage and to visitor experiences when they are inappropriate, in terms of park objectives, of large scale or are located improperly.

While tourism is a market-driven industry, the management of national parks and protected areas is determined by legislative mandates

National parks and protected areas are generally established through either a legislative process or administrative decree by the government. As such, they have a mandate to provide and protect values of interest to the public. While there has been a great deal of discussion about adopting business models for their management, the first priority of park management is to protect the values legally mandated for them. Tourism, then, is

Fig. 5.1. The ongoing eruption of the Mauna Loa volcano in Hawaii creates hazardous conditions. Park managers must deal with rapidly changing hazards that affect the park's visitors. All parks must plan for some level of hazard and prepare risk management strategies. Warning sign in Hawaii Volcanoes National Park, USA. (Photographed by Paul F.J. Eagles.)

accomplished within the administrative discretion permitted by these mandates. Such tourism development, to be successful within a business model, must be responsive to the demands of the market in order to be profitable, but the responses (e.g. promotion, facilities, service quality) developed are bounded by legal and policy directives established for the protected area. Such responses include increased customer service, appropriately designed promotional strategies, facilities and food service, and visitor research. Managers have a responsibility within the context of the park mandate to provide high-quality recreational opportunities.

Negative impacts from visitor use follow predictable patterns that can be used to structure management systems and actions

We have noted the universal concern about the negative environmental impacts of tourism development on parks and protected areas. The relationship between use levels and impacts are somewhat complex but can be schematically depicted in a general sense, although the specific values are only determined on a site-by-site basis. Nevertheless, the generalized curve, shown in Fig. 5.2, demonstrates that negative environmental and social impacts occur at relatively low use levels, increase rapidly then, at some point, increase only gradually. In Fig. 5.2, three potential relationships between use level and negative impacts are shown. Curve A characterizes the relationships generally uncovered through research on relationships between recreation use levels and negative impacts and indicates that relatively large amounts of impacts occur with relatively small amounts of use (Leung and Marion, 2000; Leung *et al.*, 2001). It suggests that factors other than use level, such as visitor behaviour, timing and season of use, and type of soil, vegetation or animals affected are also important in determining levels of impact.

This curve carries at least three implications for visitor management: (i) limiting use will probably be ineffective in controlling impacts except at very low levels of use; (ii) since use is not strongly related to impact over most of the range of the curve, other techniques to manage impacts will need to be implemented; and (iii) since there is no carrying capacity, managers must determine how much impact is acceptable. Judgements of acceptability are ultimately value statements, which can incorporate science, but science itself cannot identify the level of acceptability.

Figure 5.3 shows the relationship between the amount of visitor-induced impact and distance from the source of impact, particularly concerning impacts on wildlife from human use of a trail. The impact is generally highest closest to the trail, but as distance from the trail increases, the impact drops quickly at first, then more slowly. As with the more generalized use–impact relationship shown in Fig. 5.2, park and protected area managers must still determine how much impact is acceptable. The relationships shown in Figs 5.2 and 5.3 show only the amount

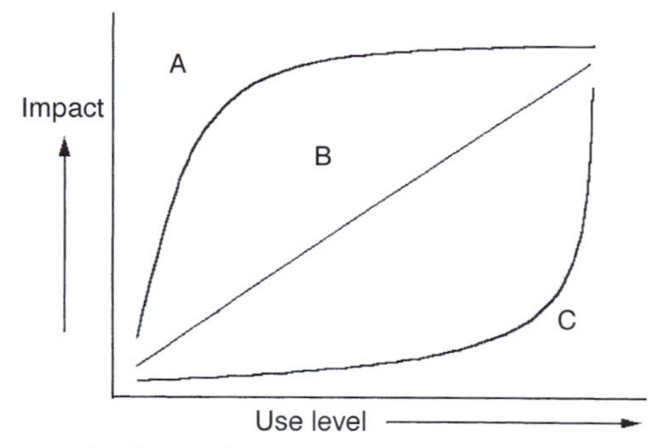

Fig. 5.2. Three potential relationships between use levels and amount of resulting biophysical and social impact. Curve A represents a situation where impacts increase rapidly with small amounts of use, and then the rate of increase decreases as use level rises. In this situation, there is also no intrinsic landscape carrying capacity. Settings characterized by even moderate levels of use would have to experience significant reductions in order to reduce impacts. In many cases, such reductions would still have little effect on the level of impact. Curve B represents a situation where impacts are a linear function of use level. In this situation as use increases, impacts increase in some linear proportion. Management strives to develop the coefficient linking use and impact (again through research). While impacts can be predicted in a relatively straightforward way, management is still left with the problem of determining an acceptable level of impact. Curve C represents a situation where the level of impact increases relatively gradually to a particular region of the curve and then begins to accelerate rapidly. If this relationship was to hold, landscapes could be characterized as containing an intrinsic carrying capacity. Management would only have to uncover this region of the curve (through accepted research protocols) and then limit use accordingly. Research in both biophysical and social impacts indicates that Curve A represents the nature of the relationship between use and impact.

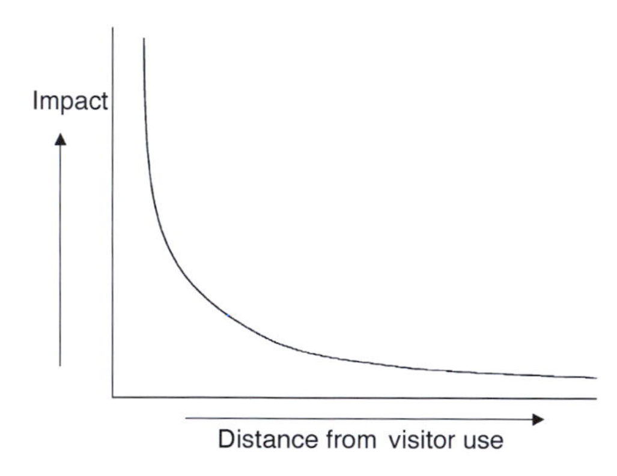

Fig. 5.3. The relationship between the amount of negative, human-induced impact and the distance from the human activity. Impacts decrease rapidly as distance increases, then more slowly. While this graphic shows how much impact may occur, it does not indicate how much negative impact may be acceptable, given the presence of recreational use.

of impact; the question of acceptability is a separate decision, which again reflects the values of those making the decision. Once decisions on the acceptability of negative impacts are made, visitor management actions, such as restrictions on group size, use of buffer zones, type of use and so on, can be made, implemented and enforced.

Providing Diversity of Recreation Opportunities: the Key to Quality Recreation and Park-based Tourism

Park managers are confronted with a difficult and challenging task: providing high-quality recreation opportunities attractive to tourists while protecting the values on which those opportunities are based. Since preferences for recreation opportunities vary over time and over a population of people, diversity in preferences exists. The principle of diversity has long been established in the recreation management literature, most notably with Wagar's classic article 'Quality in outdoor recreation' (Wagar, 1966). In this article, Wagar depicted schematically visitor preferences for development, ranging from primitive to highly developed (see Fig. 5.4). Some visitors prefer few developments during their visits; others prefer more development. Development that is aimed at the 'middle ground' will 'miss' the preferences of most visitors, thus leading to unsatisfactory experiences.

Wagar's early conception has been reformulated as the notion of a recreation opportunity spectrum (ROS) (Clark and Stankey, 1979; Driver *et al.*, 1987). The ROS focuses on three types of characteristics of a setting:

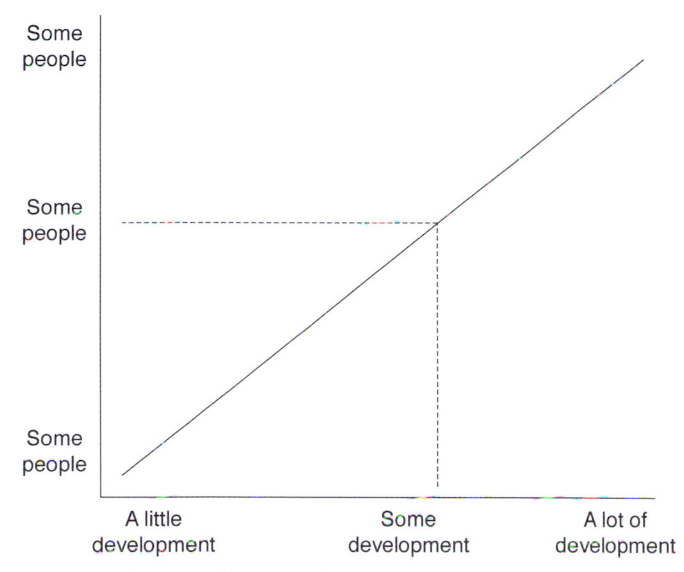

Fig. 5.4. Tourist preferences for recreational setting attributes vary over a population, as shown schematically here for facilities and developments. Some people prefer lots of facilities, while others prefer few. Planning that does not recognize this variation (diversity) and aims for a mythical 'average' visitor will lead to homogenized experiences and result in dissatisfaction on the part of most visitors. Over a particular park, and over a region, planners should identify the diversity of experiences visitors seek and, within the mandates for the park, plan for this diversity. Adapted from Wagar (1966).

biophysical attributes, managerial attributes and social attributes. The value of each of these attributes varies along a continuum.

Biophysical attributes deal with the amount, location and visibility of human-induced changes in the natural environment, for example, visitor facilities such as roads, trails, visitor centres and the presence of other resource uses such as grazing, farming, fences and so on. The amount and visibility of these facilities and uses vary on a continuum of how much modification from natural environmental settings they induce; this continuum ranges from not being present to dominating the nearby environment. The visitor's experience varies depending on where on this continuum they are currently located and their expectations.

Managerial attributes may also vary, depending on the amount and type of regulations placed on visitor behaviour and the visibility of managers, rangers, naturalists and law enforcement personnel. Again, these attributes can be placed on a continuum of not being present at all to being present, highly visible or intrusive almost continually.

Finally, social setting attributes, which include the types of visitors, their density and behaviours may also vary. For example, one area of a park may be typified by a large number of visitors that are continuously present while in other areas there may be only a few visitors observed only sporadically.

These three types of attributes may be placed on a continuum, the recreation opportunity spectrum (ROS), as shown in Fig. 5.5. Since this is a continuum, theoretically there are an infinite number of points. More practically, managers may identify 4–10 major classes of settings; one example is shown in Fig. 5.5. Each descriptive setting provides a somewhat different combination of attributes. Such combinations lead to opportunities for visitors to achieve certain recreational experiences, but they do not determine what experiences visitors create. Recreational experiences are created by visitors out of the attributes they experience and combine in their heads; what managers do is enhance the probability that a set of experiences may occur or will not occur.

Each ROS setting class involves a description of managerial, social and biophysical setting attributes. An example of such descriptions is shown in Box 5.1 (pp. 106–107). While these descriptions may be viewed as rather vague, they are relative to each other and thus provide some overall direction for zoning a protected area. These opportunity settings are not recreation activities, they are the places where such activities occur. Thus, backpacking can occur across the spectrum, but would most likely happen towards the more primitive end of the spectrum. In some cases, the nature of the setting class does restrict the type of activity permitted. For example, motorized recreation generally would not be found in primitive or semi-primitive recreational settings because such activity is inconsistent with the notion of a quiet, non-mechanized and pristine setting.

The character of a landscape type – mountains, valleys, deserts, marshes, for example – is generally not a part of an ROS description.

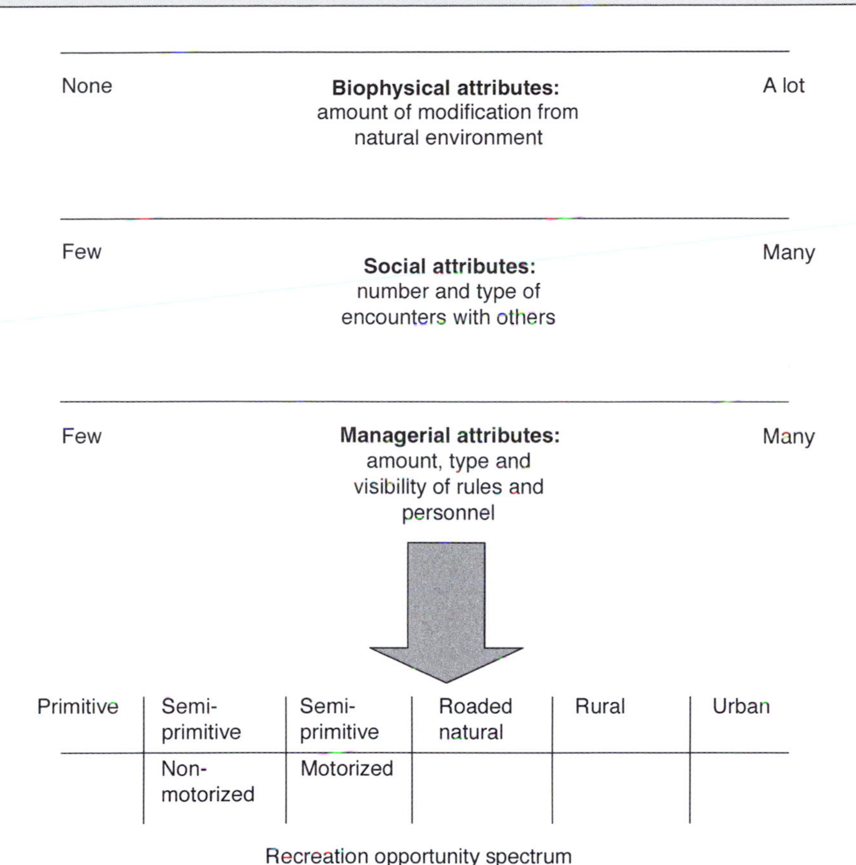

Fig. 5.5. The combination of biophysical, social and managerial attributes gives rise to the recreation opportunity spectrum, the diversity of setting opportunities that may exist at a park or in a region or that may be demanded by a population. The spectrum is divided into 'classes' to make it easier for management. Here, the classes shown are those used by the US Forest Service.

Landscapes are a dimension different from recreational settings in that opportunities can be found or allocated to different landscapes. For example, a primitive recreational opportunity can occur in an alpine area or in a desert. A highly developed recreation opportunity likewise can be found across landscape types.

Thus, over space, quality in recreation opportunity is assured by identifying places where various types of recreation opportunities can be or are provided. Within the objectives established for a park, managers may provide for this diversity in demand by identifying the elements of settings that may be varied to meet such preferences. Through the use of information and other appropriate management actions, managers influence the location of visitor activities so there is a closer match between expectations and the setting.

Box 5.1. Typical example of recreation opportunity spectrum class descriptions. These were developed for the Grand Staircase-Escalante National Monument in the state of Utah, USA. (Source: Bureau of Land Management, The Grand Staircase-Escalante National Monument Draft Environmental Impact Statement; www.ut.blm.gov/monument/Monument_Management/Initial%20Planning/deis/appendices/a20_A.html)

ROS class
Physical setting
Social setting
Managerial setting

Opportunity Class I (Primitive)
Area is characterized by essentially unmodified natural environment of fairly large size.
Concentration of users is very low and evidence of other users is minimal.
Only facilities essential for resource protection are used. No facilities for comfort or convenience of the user are provided.
Spacing of groups is informal and dispersed to minimize contacts between groups.
Motorized use within the area is not permitted.

Opportunity Class II (Semi-Primitive Non-Motorized)
Area is characterized by a predominantly unmodified natural environment of moderate to large size.
Concentration of users is low, but often other area users are evident.
Facilities are provided for the protection of resource values and the safety of users.
On-site controls and restrictions may be present but are subtle.
Spacing of groups may be formalized to disperse use and limit contacts between groups.
Motorized use is not generally permitted.

Opportunity Class III (Semi-Primitive Motorized)
Same as Semi-Primitive Non-Motorized, except that motorized use is permitted.

Opportunity Class IV (Roaded Natural)
Area is characterized by a generally natural environment. Resource modification and utilization practices are evident, but harmonize with the natural environment.
Concentration of users is low to moderate. Moderate evidence of the sights and sounds of humans.
On-site controls and restrictions offer a sense of security.
Rustic facilities are provided for user convenience as well as for safety and resource protection. Facilities are sometimes provided for group activity. Conventional motorized use is provided for in construction standards and design of facilities.

Opportunity Class V (Rural)
Area is characterized by a substantially modified natural environment. Resource modification and utilization practices are evident.
Concentration of users is often moderate to high. The sights and sound of humans are readily evident.

Continued

Box 5.1. *Continued.*

A considerable number of facilities are designed for use by large numbers of people. Facilities are often provided for specific activities. Developed sites, roads and trails are designed for moderate to high use. Moderate densities are provided far away from developed sites.
Facilities for intensive motorized use are available.

While ROS was originally designed for multiple use lands in the USA, it has been applied to national park frontcountry and backcountry settings in a variety of countries. The concept of ROS may be applied with creativity to other settings. The notion of an opportunity setting may apply to cultural sites, where there is the ability to vary the type of educational and cultural experience offered. In some cases, settings – particularly the managerial setting – may be varied across time, such as days of the week, in order to provide different types of opportunities. For example, managers of a protected river could allow many floaters (boaters, kayakers, etc.) on a few days of the week and a few floaters on other days, thus permitting opportunity for solitude on at least some days.

The ROS thus provides a framework not only for assessing what recreational opportunities may be provided, but also to determine the impacts of other management actions – such as road construction and facility development – on the existing supply of opportunities.

Park Zoning as a Management Tool

Biophysical and social conditions in national parks and protected areas often vary from one place to another. Some places are heavily visited, while others receive almost no visitors. Facilities, roads and modification of the natural environment are often not spread uniformly across the park landscape but are more likely to be highly concentrated. Visitors expect different conditions in different regions of a park. This naturally occurring diversity represents a *de facto* zoning of park visitor opportunities and heavily influences management actions. This diversity may or may not be desirable. The primary approach to managing and protecting this diversity is through the use of explicit zoning strategies.

Zoning is not only a method of providing appropriate locations for desired or preferred recreation opportunity settings, but also a tool to direct and control the spread of visitor-induced impacts to previously determined levels (Haas *et al.*, 1987). Zoning involves the allocation of differing recreation opportunities, biophysical conditions and management actions to different places within a park. Each zone represents a different land use or recreational opportunity. Land use zoning is commonly used in urban areas to ensure that incompatible developments (such as residential and heavy industrial uses) are not juxtaposed; such reasoning is appropriate also for parks and protected areas. In park and protected area planning, zoning is often based on the types of recreation settings managers wish to provide, thus linking back to the ROS concept, and on attempts to control visitor impacts. Zoning can be based on the types of activities permitted, such as buildings and facilities, roads or mechanized recreational uses. It may also be based on the amount of impacts determined to be acceptable in different areas of a park.

Zoning is a prescriptive management act; it is based on current conditions (description), management objectives for the park, the location of important values and features, demand for recreational opportunities and tourism services, and regional supplies of resources, opportunities and features. Since zoning is a prescriptive act, allocation of zones may result in changes to existing land uses and recreation opportunities. Use of concepts such as ROS allows managers to assess the impacts on supplies of different opportunities.

Each zone developed for a park or protected area will have a description of the desired and acceptable conditions. The description may include a list of management actions acceptable or desirable (e.g. group size limits, restrictions on vehicles) within the zone. When used in conjunction with the Limits of Acceptable Change process described later in this chapter it will also include quantitative standards of acceptable conditions. There is no set number of zones that a park should contain. The number and description of zones is situation specific. However, there is a limit to the capacity of humans to deal with different types of zones. In a park or protected area, managers should think carefully about having more than ten different types of zones, while most areas will have at least three zones.

Figure 5.6 shows an example of zoning used in parks and protected areas. Box 5.2 shows definitions of zones as used in the Philippines' protected area system.

Criteria Needed to Judge Visitor Management Actions

Managers have a repertoire of tools with which to manage tourism and visitors to national parks and protected areas. The selection of any given tool will be influenced first by any specific legislative mandate that may state whether a particular type of use is to be permitted, and then by a set of criteria (Checkland and Scholes, 1990) that can be generally stated as follows:

- *Efficiency*: the smallest amount of financial and personnel commitment to achieve a given end. Efficiency is important because budgets are limited; managers will want to stretch funding to cover as much management as possible.

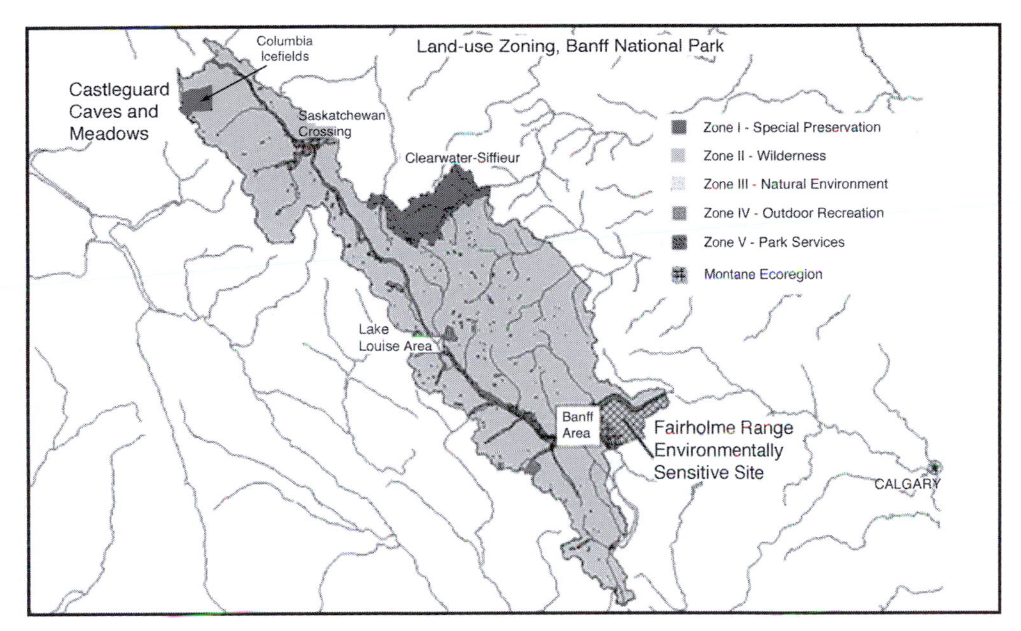

Fig. 5.6. Land-use zoning in Banff National Park, Canada. The park uses six zones; each zone provides opportunities for somewhat different recreation experiences and allows different levels of impact.

Box 5.2. Management zoning as used in the National Integrated Protected Areas System of the Philippines. (Source: Department of Environment and Natural Resources Department Administrative Order No. 25, Series of 1992; sunsite.nus.edu.sg/apcel/dbase/filipino/regs/phrnip.html)

a. *Strict Protection Zone*: Areas with high biodiversity value, which shall be closed to all human activity except for scientific studies and/or ceremonial or religious use by indigenous communities.

b. *Sustainable Use Zone*: Natural areas where the habitat and its associated biodiversity shall be conserved but where consistent with the management plan and with Park Area Management Board (PAMB) approval:

i. Indigenous community members and/or tenured migrants and/or buffer zone residents may be allowed to collect and utilize natural resources using traditional sustainable methods that are not in conflict with biodiversity conservation requirements;

ii. Research, including the reintroduction of indigenous species, may be undertaken; and

iii. Park visitors may be allowed limited use. Provided no clearing, farming, settlement, commercial utilization or other activities detrimental to biodiversity conservation shall be undertaken. The level of allowable activity can be expected to vary from one situation to another.

c. *Restoration Zone*: Areas of degraded habitat where the long-term goal will be to restore natural habitat with its associated biodiversity and to rezone the area to a

Continued

Box 5.2 *Continued.*

more strict protection level. Initially, natural regeneration will be assisted through such human interventions as fire control, cogon grass suppression and the planting of native species including indigenous pioneer tree species as well as climax species. Exotic species (not native to the site) shall not be used in the restoration process. Existing houses and agricultural developments may be allowed to remain initially but would be phased out eventually.

d. *Habitat Management Zones*: Areas with significant habitat and species values where management practices are periodically required to maintain specific non-climax habitat types or conditions required by rare, threatened or endangered species. Examples would be forest openings for the tamaraw or brushy forest for the Philippine tarsier. Human habitation and sustainable use may be allowed if they play a habitat management role.

e. *Multiple-use Zones*: Areas where settlement, traditional and/or sustainable land use, including agriculture, agroforestry, extraction activities and other income-generating or livelihood activities, may be allowed to the extent prescribed in the management plan. Land tenure may be granted to tenured residents, whether indigenous cultural community members or migrants.

f. *Buffer Zone*: Areas outside the protected area but adjoining it that are established by law (Section 8 of the Act) and under the control of the Department of Environment and Natural Resources (DENR) through PAMB. These are effectively multiple-use zones that are to be managed to provide a social fence to prevent encroachment into the protected area by outsiders. Land tenure may be granted to occupants who qualify. Buffer zones should be treated as an integral part of the protected area in management planning.

g. *Cultural Zones*: Areas with significant cultural, religious, spiritual or anthropological values where traditional rights exist and ceremonies and/or cultural practices take place.

h. *Recreational Zones*: Areas of high recreational, tourism, educational or environmental awareness values where sustainable ecotourism, recreational, conservation education or public awareness activities may be allowed as prescribed in the management plan.

i. *Special Use Zones*: Areas containing existing installations of national significance, such as telecommunication facilities, irrigation canals or electric power lines. Such installations may be retained subject to mutual agreements among the concerned parties, provided such installation will not violate any of the prohibitions contained in Section 20 of the Act.

- *Effectiveness*: the extent to which a technique actually works. For example, if managers implement a use limit policy to reduce biophysical impacts, such limits must reduce the impacts. To determine effectiveness, a monitoring programme will be needed.
- *Efficacy*: how a technique may contribute to or detract from the larger goals of park management. Parks and protected areas, for example, are used by visitors during their free time; restrictions on their behaviour may not be efficacious because such actions detract from the larger societal goal of providing high-quality recreation opportunities.

- *Equity*: the distributional effects of a particular technique. Managers will want to know who is affected by an action and how. In the USA, for example, an environmental justice presidential executive order requires agencies to determine whether federal actions will have an unnecessary and unfair impact on minorities.

Not all proposed actions will meet each evaluative criterion equally; in some cases there may be distinct and difficult trade-offs involved among the criteria. Public participation will help managers to determine how important each criterion is in each decision situation. What is important in developing proposed actions is that managers establish a comprehensive, systematic visitor management programme, such that policies are integrated with one another, and that monitoring is an intrinsic component of implementation.

Visitor Management Systems

Development of visitor management programmes is directed not only by the park and park management agency mandate, but also frequently by other legislation that may have a bearing on a particular park or how the government conducts planning and decision-making processes. An example of such legislation may be the national level environmental assessment laws in a particular nation. Other legislation may dictate that park agency planning follows specific steps or contains specific components.

Yet, within these constraints lie a variety of approaches to visitor management planning that tend to: (i) make explicit various values and assumptions about what visitors want and parks can provide; (ii) provide a framework for defining and framing the challenge of visitor management; (iii) require definitive statements of objectives and desired conditions; and (iv) force decisions on the use of specific management tools to be made following decisions on objectives.

Visitor management is an administrative action oriented towards maintaining the quality of park resources and visitor experiences. In this sense, managers act to maintain the values established in the organic legislation establishing the park or recreation area and, within the discretion allowed by this mandate, provide high-quality opportunities for visitors to experience those values. In many, but not all situations, management tends to focus on the negative impacts resulting from unrestrained visitor activity. In other situations, management acts assertively to create and maintain opportunities for visitors to view, experience, learn about and appreciate their natural and cultural heritage. In many cases, park staff will have developed assertive service quality programmes to ensure that visitors can access high-quality recreational and learning opportunities. While the management systems that have been developed to assist managers in maintaining natural and cultural heritage tend to focus on the

Fig. 5.7. All parks require a cadre of staff that are educated in all aspects of park management and dedicated to the ideals of the park. Dan Strickland, Chief Park Naturalist, Algonquin Provincial Park, Canada. (Photographed by Paul F.J. Eagles.)

negative human-induced impacts, they can just as well be used to create positive impacts to both resources and people.

A variety of visitor management systems exist. These include the Limits of Acceptable Change (LAC) planning system, the Tourism Optimization Management Model (TOMM), Visitor Impact Management (VIM) planning, the Visitor Experience and Resource Protection process (VERP), Visitor Activity Management Planning (VAMP), and the concept of the recreation opportunity spectrum discussed earlier and the derivative tourism opportunity spectrum (Stankey *et al.*, 1985; Kuss *et al.*, 1990; Manidis Roberts Consultants, 1997; McCool and Cole, 1997; US Department of the Interior, 1997; Nilsen and Grant, 1998).

While these systems all have different names, they have a number of components in common. Each of these systems has a similar theoretical foundation and contains comparable steps and processes. All require initial statements of park values and objectives. Each deals with understanding the issues and conflicts present in a park. The LAC, VIM, VERP and TOMM systems include requirements for identifying quantitative indicators and standards to determine how much human-induced change is desirable. Most require early and continuous public involvement to be successful. Each of these systems was designed to address visitor management questions as a result of failures of approaches designed to establish numerical recreational carrying capacities (see McCool and Lime (2001) for a critique of carrying capacity).

In this chapter, we focus on the Limits of Acceptable Change system, initially developed for USA Forest Service administered wilderness areas and now used around the world in a variety of parks and protected areas. We will also present the Visitor Activity Management Planning process developed in Canada which is a somewhat different approach. These two processes have been extensively used and tested in different situations, and they are broadly representative of contemporary visitor management planning approaches.

The Limits of Acceptable Change planning process

The Limits of Acceptable Change system was developed in response to dissatisfaction with attempts to establish a numerical recreational carrying capacity for wilderness and whitewater rivers. The carrying capacity concept often assumed that the intrinsic character of the land base would determine how many people were too many. However, this approach made a number of assumptions that have proved to be highly questionable and, in addition, took managers down a path that in the short term seemed reasonable, but in the long term did not solve the problems with which they were confronted (McCool and Lime, 2001). The LAC process reframes the question of 'how many visitors are too many?' to one of 'what are the appropriate/acceptable biophysical and social conditions in a park/wilderness?' This changes the discussion from one emphasizing the negative impacts of human activity to one of focusing on the desired outputs of management and then determining the best actions (inputs) to get there (Stankey and McCool, 1984; McCool and Cole, 1997).

LAC is a process for defining the biophysical and social conditions that are acceptable or desirable at different locations in protected areas, and then determining the management actions and approaches most appropriate for enhancing, maintaining or restoring those conditions. Biophysical conditions refer to the amount of change from natural conditions caused by park tourism and associated developments. Social conditions refer to the number, frequency and type of individuals a visitor may encounter during a typical visit. Determining the appropriateness of these conditions and how to manage for them is best accomplished through a process that is systematic, explicit, defensible and rational, coupled with extensive public participation. In this section, we briefly present the elements of the LAC process as they may apply to managing visitors in parks and protected areas.

The key elements in the LAC process are to: (i) identify the acceptability of biophysical and social impacts; (ii) develop management techniques to ensure that the standard of acceptability is not breached; (iii) monitor resulting conditions; and (iv) change management techniques if needed.

Since acceptability is determined by cultural norms, this means the process to identify standards must be both explicit and open to participation by park stakeholders. Simply defined, LAC is a process for determining the resource or social conditions that are acceptable, and then prescribing a set of management actions to achieve those conditions. The LAC system comprises nine steps. While some of the individual elements of LAC had been applied in the early 1980s in several wilderness areas, the first complete application came in the Bob Marshall Wilderness Complex in northwestern USA.

The LAC process recognizes the inevitable impacts resulting from recreational use. Park and protected area managers might want to retain pristine conditions throughout an area, but the reality is that, once use occurs, biophysical conditions change; soils become compacted, vegetation suffers mechanical damage and is disturbed, duff is removed from campsites. Social conditions also change; intergroup encounters increase, conflicts arise. The nature and extent of these changes will vary throughout a protected area because of differences in types and amounts of use, sensitivity of the vegetation and soils, and other factors. Managers may also want to create opportunities for recreation and learning in certain places within a protected area and other opportunities elsewhere. Thus, there will be diversity in biophysical and social conditions that the LAC process explicitly recognizes. The LAC process then requires that this diversity is examined and a decision about how it should be preserved or modified is made.

To implement the LAC process, managers proceed through nine steps (see Fig. 5.8). Four comments need to be made about the nine-step process. First, the social–political context and the nature of the resource vary considerably from one country to the next, and from one park to another. Thus, with good reason, the LAC process can be modified, both in the number of steps and the order in which they proceed. However, this should only be done if there is good reason and if the manager understands the rationale for each step and its place in the LAC process.

Secondly, one does not necessarily proceed through the LAC process in a linear, unidirectional manner. As one goes through the process, new issues arise that may not have occurred to planners in earlier steps. Therefore, the substance of earlier steps may have to be modified to deal with these new issues. For example, in step 2, the number and description of prescriptive management zones are identified. A manager may feel that three zones are appropriate. However, the manager may find that in step 6, where allocations for zones are actually made, four classes are more appropriate. Therefore, step 2, and the ones following it should be modified appropriately.

Thirdly, public involvement in the LAC process is essential to its success (Fig. 5.9). The LAC process encourages explicit treatment of the many subjective and prescriptive decisions that are required to protect park values and resources. Because of these value judgements, it is important that managers encourage discussion of park planning issues within the context of many different value systems. This will allow the full range of issues and concerns to surface. These issues can be debated and many resolved early in the planning process.

Fourthly, LAC may be implemented as an initial component of a park plan or it may be conducted as a separate process tiered to a more comprehensive plan. LAC is primarily a visitor management process that does not necessarily include other issues, such as fire and wildlife

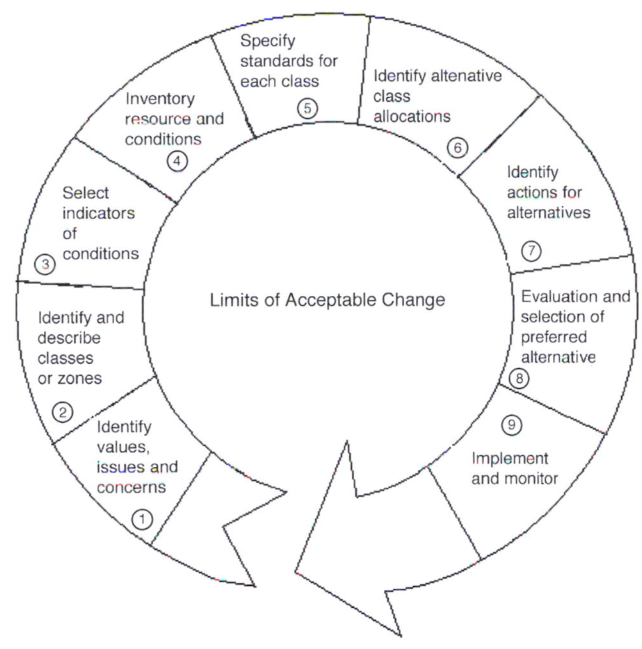

Fig. 5.8. The Limits of Acceptable Change planning process depicted as a continuous circle of planning. Each step is sequenced to build on previous steps and provide information for future steps. The process is shown as circular to indicate the importance of monitoring data feeding back into the management/planning process (adapted from Stankey *et al.*, 1985).

Fig. 5.9. Successful protected area planning in messy situations involves implementing a scientifically sound technical planning process and fully involving the affected public in the process in order to achieve a consensus about a desired future and the means to achieve it.

management. Because the LAC process requires different types and a wider array of information than traditional plans, park managers may wish to forgo some components of the LAC process when they develop an initial park plan. However, this will require that the initial plan is carefully crafted to allow incorporation of LAC components later.

Step 1: identify special values, issues and concerns attributed to the area
Citizens and managers meet to identify what special features or qualities within the area require attention, what management problems or concerns have to be dealt with, what issues the public considers important in the area's management and what role the area plays in both a regional and a national context. This step would normally begin with a review of legislative mandates and current policies. These are then translated into specific objectives. (Objectives are statements of desired conditions.) Issues function as barriers to achieving those conditions. Special values may be recognized in enabling legislation or in governmental policy statements. The recognition of these values may come from new research or from visitor response to the area. Many of these values may be located only in a portion of the area, such as a cave, waterfall, prehistoric human occupation site or vegetation type.

This step encourages a better understanding of the cultural values and natural resource base, such as spiritual meanings, the presence of historically significant events and the sensitivity of natural environments

Fig. 5.10. What are the appropriate/acceptable biophysical and social conditions in a park/wilderness? The decision was made in the Maasai Mara Game Reserve in Kenya not to construct roads. Consequently, the lack of roads forces the safari vehicles to travel cross-country, resulting in soil and vegetation damage. Interestingly, Serengeti National Park in Tanzania, in the same nearby ecosystem, constructed all-weather roads for safari vehicles. Safari vehicle tracks in Maasai Mara Game Reserve, Kenya. (Photographed by Paul F.J. Eagles.)

to recreation use and tourism development. This step provides a general concept of how the resource could be managed and an understanding of the principal management issues. In this step, the interaction of the area with the local tourism industry is also identified. Understanding objectives, values and issues is important as these lay the foundation for much of what will follow in the LAC process.

Step 2: identify and describe recreation opportunity classes or zones Most park and protected area settings of sufficient size contain a diversity of biophysical features. For example, a marine protected area may contain reefs, underwater cliffs and corals. Likewise, social conditions, such as level and type of use, amount, density and type of development, and types of recreation experiences, vary from place to place. The type of management needed may vary throughout the area. Opportunity classes describe subdivisions or zones of the natural resource where different social, biophysical or managerial conditions will be maintained (see previous sections on ROS and zoning). For example, in a marine park, deeper reef settings will require the use of scuba gear for all users while in shallower areas snorkels may be adequate. The shallower areas may also show more negative impact from human use, such as effects on coral, than deeper areas.

The classes that are developed represent a way of defining a range of diverse conditions within the park setting. And, while diversity is the objective here, it is important to point out that the conditions found in all cases must be consistent with the objectives laid out in the area's organic legislation or decree. In this step, the number of classes is defined as well as their general biophysical, social and managerial conditions. This step focuses solely on identifying the range and classifying this range into a number of classes, but does not determine where these classes may eventually be located.

Zoning a park or protected area in this way serves several functions. First, it allows planners, managers and visitors to agree more explicitly on the amount of human-induced change that may be permitted in any given area. Secondly, it is a tool to influence where developments and use patterns may or may not be challenged. Thirdly, it suggests what the diversity of opportunities will be in the park. At this point, however, the zones are only described, not actually allocated. Allocation must wait until more information about the condition of the park is established.

Step 3: select indicators of biophysical and social conditions Indicators are specific elements of the biophysical or social setting selected to be indicative of the conditions deemed appropriate and acceptable in each opportunity class. Because it is impossible to measure the condition of and change in every biophysical or social feature within a protected area setting, a few indicators are selected as measures of overall health, just as we periodically monitor our blood pressure rather than undergo frequent comprehensive tests of blood chemistry. Indicators should be easy to measure quantitatively and relate to the conditions specified by the opportunity classes, and should reflect changes in recreational use.

Indicators are an essential part of the LAC framework because their value reflects the overall condition found throughout an opportunity class or zone. It is important to understand that an individual indicator measured alone might not adequately depict the condition of a particular area. It is the bundle of indicators that is used to monitor conditions. An example of an indicator of impact on vegetation in a terrestrial park might be the amount of barren soil created by campsites, while in a marine setting, visibility in metres may be a useful and appropriate indicator of water quality.

What are characteristics of good indicators? How does one choose what indicators to measure? The choice of indicators is greatly influenced by the issues and concerns identified in step 1 of the LAC process. For example, if solitude is identified as an issue, then it would be logical to select intergroup encounters as an indicator. Indicators that are subject to management influence are also important to select. For example, trail conditions (erosion, trail width, presence of boggy spots, multiple parallel trails in meadows, etc.) in backcountry areas are frequently identified as major management issues. What would be an appropriate indicator for trail conditions? In backcountry areas where this question has been considered, no good indicator has been identified although a number have been proposed. The primary reason is that trail conditions tend to be more a function of location, construction technique and maintenance than use level or visitor behaviour, the primary orientation of many management actions. Generally, it has been found to be more efficient to deal with the issue of trails with a management policy.

In other cases, where management and the public are unwilling to make trade-offs between impacts and recreational access, indicators would not necessarily be useful. For example, at cultural resources such as historic sites, management has determined that no impacts on the cultural attributes will be permitted. In this case, while an indicator might be useful to determine if impacts have occurred, management generally will restrict recreational access rather than permit impact.

Beyond these types of issues, there are several criteria useful in evaluating indicators (Merigliano, 1989; Martin, 1990):

1. *Sensitivity to change.* Is the proposed indicator sensitive to human uses and behaviours? That is, will the indicator reflect impacts that may occur? For example, water quality is often an issue and faecal coliform (a measure of human faecal contamination) count is sometimes suggested as an indicator. However, a variety of research shows that there is little correlation between visitor use levels and faecal coliform counts in backcountry situations.

2. *Accepted definition.* Is there a universally accepted definition of the indicator? For example, the definition of blood pressure is the same whether a doctor practises in New York or Paris. A common proposed indicator is intergroup encounters. What does this mean? Does meeting the same group on the trail twice in one day mean two encounters or one? The point is, the indicator should have an unambiguous definition.

3. *Consistent measurement methodology*. Closely related to the second criterion is the need to have an indicator that can be measured reliably by different people. Reliability means that each observer measures the indicator in the same way. Ideally, the measurements should also be accurate, which means they are valid, on target.

4. *Simple and easy to implement*. Indicators should also be designed so that those measuring them do not require extensive training. Indicators should be easy enough to measure so that they do not drive up the costs of management.

5. *Reflect overall conditions*. The indicators chosen should, as a group, be representative of overall conditions. Indicators that are too narrowly focused may not be helpful in assessing the overall status of human impacts in the wilderness.

6. *Quantifiable*. Indicators should be defined so that they are directly measurable on an interval or ratio level scale. Indicators that are classifications or judgements (e.g. 'crowded', 'heavily impacted') should be avoided. Indicators should carry clear implications for measurement.

Some example indicators are shown in Table 5.1.

Step 4: inventory existing resource and social conditions Inventories of biophysical and social conditions can be time consuming and expensive components of planning; indeed they usually are. In the LAC process, the inventory is guided by the indicators selected in step 3 and does not occur until managers and the public understand issues, values and objectives. This understanding provides the context needed to determine what should be inventoried. For example, level and type of development, use

Table 5.1. Examples of indicators of biophysical and social conditions related to park tourism use.

Biophysical conditions
Forage utilization
Number of backcountry campsites per km^2
Recreation facility 'footprint' in ha
Number of damaged trees at campsite
Amount of barren soil at campsite
Social conditions
Number of campsites occupied within sight or sound
Number of other people encountered on trail
Location of encounters
Stress produced from encounters with others

density and human-induced impacts on coral in a marine park might be measured, but that information is only useful if there are objectives or issues concerning levels of impact or amount of development. Other variables, such as location of different corals, shipwrecks, docks and mooring spots, can also be inventoried to develop a better understanding of area constraints and opportunities. Such inventory information will be helpful later when evaluating the consequences of alternatives. Inventory data are mapped so both the condition and location of the indicators are known. The inventory also helps managers to establish realistic and attainable standards. By placing the inventory as step 4 rather than step 1, as is usually recommended, planners avoid unnecessary data collection and ensure that the data collected are salient and will actually be used.

Step 5: specify standards for resource and social conditions in each opportunity class In this step, the range of conditions for each indicator considered desirable or acceptable across opportunity classes is identified. By defining those conditions in measurable terms, managers provide the basis for establishing a distinctive and diverse range of park or protected area settings. Standards serve to define the limits of change acceptable to the public and managers. In general, standards represent the maximum permissible conditions that will be allowed in a specific opportunity class. They are not necessarily objectives to be attained.

The use of such standards is not a new idea; they are often applied in many areas of human endeavour, such as water quality. A common water quality standard concerns the presence of human faecal material. The standard is generally written to allow up to a maximum faecal bacteria count, such as a count of 14 per 100 ml sample. Air quality standards are written in similar ways, such as an average maximum particulate count for a 24 h period of 150 $\mu g\ m^{-3}$. Standards thus written represent a trade-off between human health objectives and others such as economic development, costs of pollution control and so on. They suggest, in the case of water quality, that the risk of disease with counts below 14 is so small that economic development may proceed until that point. The decision to place the standard at a certain point represents a value judgement concerning the risks associated with less restrictive standards and the costs incurred to forms of human activity. In the process of making these value judgements, science is used to inform policy makers and the public of the consequences of alternative standards. However, the choice of the specific standard – the limit of acceptable change – is clearly a social and political decision informed by science but not determined by it.

The inventory data collected in step 4 play an important role in setting standards. The standards defining the range of acceptable conditions in each opportunity class should be both realistic and attainable; however, they should do more than mimic existing (and possibly unacceptable) conditions (Fig. 5.11). An example of a standard might be that no more than 75 m^2 of barren soil will be permitted at any given backcountry campsite.

Fig. 5.11. Park managers utilize high levels of staffing to educate and direct the visitors near the lava front in Hawaii Volcanoes National Park. This staffing activity is expensive but deemed necessary due to the high levels of hazards affecting the visitors. The decision to place a management standard at a certain point represents a value judgement concerning the risks associated with less restrictive standards and the costs incurred to humans and their activities. Lava entering the sea in Hawaii Volcanoes National Park, USA. (Photographed by Paul F.J. Eagles.)

In a marine setting, a standard might be written as 'Visibility in the water should always be a minimum of 15 metres at a depth of 3 metres'.

The role of standards in management is often debated. Are they 'yellow' lights, simply providing a warning of unacceptable conditions, or are they 'red' lights, indicating the maximum permissible conditions? There are advantages and disadvantages to each interpretation. However, when standards are interpreted as 'red' lights, this means that the managing agency must not allow conditions to get any worse. Therefore, the agency is committed to taking any action necessary to prevent worsening conditions. If standards are interpreted as a 'yellow' light, agencies have more discretion about what to do and when, but then there is no commitment to remedial action.

Step 6: identify alternative opportunity class allocations Most national park or protected area settings could be managed in several different ways given their legislative mandate. Indeed, parks often differ significantly in the amount of development, human density (both residents and visitors) and recreational opportunities available, not only between areas but within an area as well. In this step, we begin to identify alternatives that have a reasonable chance of meeting the objectives and preserving the values established in step 1. Using information from step 1 (area values, issues and concerns) and step 4 (inventory of existing conditions), managers and citizens can begin to jointly explore how well different opportunity class allocations meet the various competing interests, concerns and values. For example, one alternative scenario for a protected area may emphasize greater amounts of park tourism in order to capitalize on the growing interest in ecotourism activities. Another alternative may place greater emphasis on protection of pristine conditions or rare wildlife populations. These two alternatives would be expressed by different zoning or allocation of opportunity classes. Inventory information is then used to help assess the consequences of each alternative.

Step 7: identify management actions for each alternative The alternative allocations proposed in step 6 are only the first step in the process of developing a preferred alternative. In addition to the kinds of conditions that would be achieved, both managers and citizens need to know what management actions will be needed to attain the desired conditions. For example, for places where the amount of acceptable change is limited, managers may choose highly restrictive techniques, while in areas of a park where more change is permitted, facilities may be developed to manage park tourists. In other places where park tourism is encouraged, management may emphasize facilities, interpretive services and guided programmes. In a sense, this step requires an analysis of the costs, broadly defined, that will be imposed by each alternative. For example, many people may find an attractive alternative is to protect one area of a park from any development and restore to pristine condition any impacts that might exist. However, this alternative might require such a huge commitment of funds for acquisition and enforcement that the alternative might not seem as attractive.

Managers have many tools available for use in attaining desired conditions. These range from the highly restrictive ones, where access to the park is limited, to techniques based on provision of information, as discussed below. Here, the important principle is to separate decisions about what conditions will be desired from how those conditions will be achieved. Select conditions first, then determine appropriate management actions. By identifying such actions in conjunction with the public, more informed decisions about the costs and consequences of each alternative can be made.

Step 8: evaluation and selection of a preferred alternative With the costs and benefits of the various alternatives before them, managers and citizens can proceed to evaluation, and the managing authority, based on guidance from the public, can select a preferred alternative. Evaluation must take into consideration many factors and criteria (such as efficiency, effectiveness, efficacy and equity as noted earlier). Other criteria may include the responsiveness of each alternative to the issues identified in step 1, management requirements from step 7, impacts to the tourism industry and public preferences. It is important that the factors figuring into the evaluation process and their relative weight are made explicit and available for public review. Often, such an evaluation can occur using a matrix, with alternatives in the columns and evaluation criteria in rows. Effects are noted in the intersecting cells. An example is shown in Table 5.2.

Step 9: implement actions and monitor conditions With an alternative finally selected and articulated into policy by decision makers, the necessary management actions (if any) are put into effect and a monitoring programme instituted. Plans with significant ownership by those affected have the greatest chances of implementation, so public participation throughout the LAC process including this component, is imperative.

Often, an implementation plan detailing actions, costs, timetable and responsibilities will be needed to ensure timely implementation. The monitoring programme focuses on the indicators selected in step 3, and compares their condition to those identified in the standards. We note here that monitoring is integral to any management planning process, including the LAC system; it is not a step or element that can be left out. Information developed in the monitoring process is used to evaluate the success of actions. If conditions are not improving, the intensity of the management effort might need to be increased or new actions implemented. The process is continual as shown in Fig. 5.12.

We have purposely taken some time to explain the LAC process. To some, it may seem excessively complicated, time consuming and expensive. Actually, nearly every park management planning process will include these steps or elements, but many are implicit or 'hidden', and not subject to public participation or review. In some cases, prescriptive management actions are combined with descriptive activities. In terms of

Table 5.2. Example of matrix for evaluating alternatives.

	Alternative		
Criterion	Maintain pristine conditions	Emphasize tourism development	Enhance ecotourism opportunities
Impact on wildlife			
Administrative feasibility			
Impact on cultural resources			
Job opportunities			
Effect on access of local residents			

Matrix used to summarize and compare effects of three hypothetical management alternatives for a national park. The evaluation criteria are derived from important issues and goals established for the park, and from the suggestions in the text. Planners, managers, scientists and stakeholders would complete the matrix by filling in the cells, which would depict effects. The cells could depict actual impacts, or they could rate the impacts of one alternative relative to others through the use of rating numbers (1–3 for example, with 1 showing the alternative with the least negative effects for that criterion).

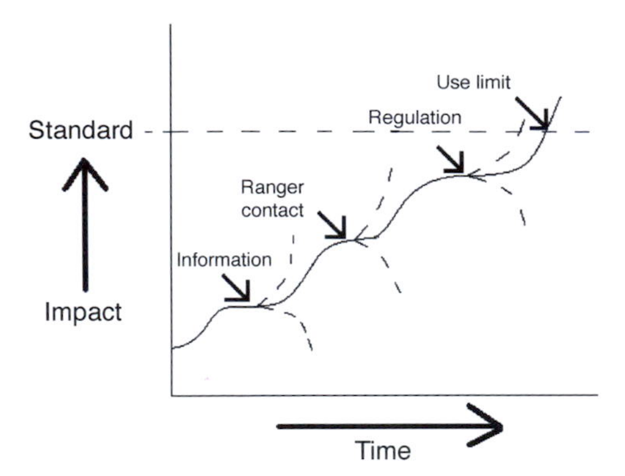

Fig. 5.12. The use of standards, monitoring and implementation of management. The straight dashed line represents the limit of impact determined acceptable: a standard. The curved solid line shows the level of impact recorded through monitoring. The curved dashed lines represent potential impact levels without management intervention. Management actions – information, ranger contact, regulation and use limit – are shown in the order in which they might be implemented. As impacts rise on the left-hand side of the graph, management implements an information programme to reduce or control impacts. This action may be effective for a time, but then impacts may rise again, triggering the use of a ranger contact method to reinforce messages in the information campaign. The action may or may not be effective; if not, a regulation may be written to further control the level of impact. Ultimately, as impacts approach the standard of acceptable change, managers may wish to implement a policy limiting the number of visitors. Alternatively, they may wish to control the spread of impacts through some type of site-hardening technique.

the public values involved, the LAC process provides managers with the explicitness, rigour and framework needed to develop defensible and understandable management actions.

The Visitor Activity Management Programme

The Visitor Activity Management Programme (VAMP) is type of planning process developed to deal with tourism in Canadian National Parks. Table 5.3 lists the steps used in the process. Note that VAMP appears to be more comprehensive in identifying the steps involved in producing a plan than the LAC process. For example, VAMP begins with the production of the terms of reference – specifications, timeframes, project personnel and so on – of the planning project. The next step in VAMP is to examine be current legislative and policy mandates and directives as well as specific values with which the park has been established. This examination forms the foundation or fence around any management actions that may occur in the planning process.

The third step in the VAMP process is to identify, organize and use existing databases that provide the needed biophysical, social and policy

Table 5.3. Visitor Activity Management Planning process (source: Nilsen and Taylor, 1997).

1. Produce a project's terms of reference.
2. Confirm existing park purpose and objectives statements.
3. Organize a database describing park ecosystems and settings, potential visitor educational and recreational opportunities, existing visitor activities and services, and the regional context.
4. Analyse the existing situation to identify heritage themes, resource capability and suitability, appropriate visitor activities, the park's role in the region and the role of the private sector.
5. Produce alternative visitor activity concepts for these settings, experiences to be supported, visitor market segments, levels of service guidelines and roles of the region and the private sector.
6. Create a park management plan, including the park's purpose and role, management objectives and guidelines, regional relationships and the role of the private sector.
7. Implementation: set priorities for park conservation and park service planning.

information. These databases can be useful in examining potential management actions, understanding the regional setting in which the park is located and in implementing policies and actions as well as understanding the consequences of proposed management regimes. In step 4, the existing management emphases are identified and evaluated in the light of biophysical capabilities, regional contexts, park values and the demand–supply relationships for park and protected area tourism. Step 5 involves a conceptual development of different ways in which the park or protected area can be managed. In this step, park visitor surveys are developed and visitor responses are segmented to better understand the preferences, attitudes and activity patterns of park visitors. In this step, the roles of the private and public sectors in providing services and recreation opportunities for visitors are identified, alternatives developed and suggested policies defined.

In step 6, the foregoing materials are put together in a physical document termed a management plan. The management plan outlines the direction and pathways for achieving the desired futures. The final component of the VAMP process is to implement the actions identified in the general management plan. In the implementation of management actions, priorities are identified, costs estimated, personnel assigned and timetables developed.

Many of the steps that occur in the LAC process are indirectly incorporated or can be included in the VAMP process if managers and planners so desire. For example, VAMP step 2 deals with park purposes, similar to what happens in LAC step 1. Step 5 of VAMP deals with producing alternative visitor activity concepts, a process similar to identifying alternative opportunity class allocations in LAC. What is important here is that managers and planners may adapt each of these processes to suit their own particular needs and to meet the needs of the social, political and biophysical context in which a particular park or protected area is located.

Case Study Number 5: Bob Marshall Wilderness (USA)

Successful Wilderness Management Using the Limits of Acceptable Change Planning System

The Bob Marshall, Great Bear and Scapegoat Wildernesses are located in north-central Montana, across the Continental Divide. The three wildernesses are immediately juxtaposed (commonly referred to as the Bob Marshall Wilderness Complex) and are managed under provisions of the 1964 Wilderness Act.[1] They comprise approximately 600,000 ha of unroaded, temperate forest, ranging from low elevation valleys to subalpine and alpine environments. Approximately 25,000 individuals visit the area each year, primarily from June to November. The June–September period is dominated by typical backpacking and horse-supported backcountry recreation trips. In the autumn, most use is for big game hunting, focused on elk. Nearly all major mammals that existed in the area prior to Euro-American occupation of North America continue today, including elk, deer, big-horn sheep, black bear and grizzly bear. Wolves may move into the area shortly, following successful restoration efforts nearby. One active backcountry airstrip exists in the area.

In 1982 the US Forest Service, which has management responsibility, embarked on a new planning effort guided by increased and continuous public participation in the planning process. Through the initiation of a task force comprising members of the public (including outfitters, environmentalists, backpackers, private horse users and pilots), scientists and managers worked cooperatively over a period of 5 years. In addition, the planning process was based on the newly developed Limits of Acceptable Change process that focused effort on addressing the question of how much change in wilderness, biophysical and social conditions is acceptable. The process followed the steps outlined above. By designing a public participation process that incorporated the full range of values involved in the wilderness, participants, including managers, developed a set of management actions that were not only effective in reducing and controlling human-induced impacts but achieved the social and political acceptability necessary for implementation. The plan was implemented in 1987.

The plan has several features significant for protected area planners. First, it establishes a series of opportunity classes or zones. These zones are designed to protect the pristine character of the wilderness, yet realistically permit some trade-offs between recreation use and human-induced impacts. There are four such zones in the area. Secondly, the plan identifies a series of indicator variables – things to monitor to ensure that conditions remain acceptable and to use to establish the effectiveness of actions implemented to control or mitigate impacts. With each indicator, explicit and quantifiable

[1] The Wilderness Act protects 628 areas in the USA from road construction, timber harvesting and a variety of other resource extraction issues. It was established primarily to ensure that not every acre in the USA would be developed. It is an act that is distinctly separate from the National Parks Act; yet national parks may contain areas of designated wilderness. Federally administered lands managed by the US Forest Service, National Park Service, Bureau of Land Management, and US Fish and Wildlife Service are eligible for designation as wilderness. Such designations must be established by an act of Congress.

standards exist. Standards are the limit to how much change from the natural baseline is acceptable in each zone, given that recreational use will occur. Thirdly, each zone has a set of management actions indicated in order of their social acceptability. The manager reaches into this tool box in the order indicated and determines what management action will be most effective and acceptable in controlling any impacts. This procedure thus encourages the use of the least intrusive management action, but allows the manager to impose use limits if no other actions work.

Recreation opportunity classes or zones in the Bob Marshall Wilderness Complex

Five zones are identified and brief summary descriptions of each zone are listed in Box 5.1 (pp. 106–107). Through the public involvement process, lands within the complex were allocated to one of the five zones. These zones then form the framework for managing human-induced impacts in the complex. Note that the zones do not carry names, just Roman numerals. In this case, proposed names were simply too confusing and did not clearly identify the types of opportunities and conditions permitted in each zone. Therefore, a decision was made to use numerals only. The names in parentheses in Box 5.1 are used here for descriptive purposes. Full descriptions of each zone are found in the plan. Each zone is described by the biophysical, social and managerial setting conditions that are acceptable. Note that some elements of the descriptions remain the same throughout the zoning, while others reflect subtle changes in acceptable conditions and management. The opportunity classes represent amounts of impact permitted on a continuum; Opportunity Class I (Primitive) is the most pristine zone while Opportunity Class IV (Roaded Natural) is the least pristine. A careful reading of the descriptions is needed to appreciate fully the differences from zone to zone.

Conclusion

Integrating tourism, parks and communities in a planning process is anything but a simple task. Good stewardship requires consideration of important values and objectives, interests that are involved and affected, and the effectiveness of various management strategies. Good stewardship begins with an understanding of the important principles and concepts that underlie visitor management. These principles, such as understanding visitor motivations and characteristics, are essential to effective implementation of various management actions.

Zoning a park is not only a way of enhancing the quality of recreation opportunities – because it reduces interactivity conflict – but also a major method of limiting the spread of unacceptable human-induced impacts. Fundamental to good practice of zoning is understanding the rationale for it as well as differentiating between the descriptive process of identifying what opportunities currently exist and the prescriptive process of determining what and where such opportunities (or others) will exist under the framework of the area's management plan.

Fundamental to planning and implementation is incorporating periodic monitoring of important biophysical and social conditions into the planning process. Monitoring – as discussed later in the book – should neither be compartmentalized into other functional areas, nor viewed as a step outside of planning. Without monitoring, we have no way of determining whether planned actions have led to their planned consequences or to 'surprises'.

References

Checkland, P. and Scholes, J. (1990) *Soft Systems Methodology in Action*. John Wiley & Sons, Chichester, UK.

Clark, R.N. and Stankey, G.H. (1979) *Recreation Opportunity Spectrum: a Framework for Planning, Management and Research*. USDA Forest Service, Pacific Northwest Forest and Range Experiment Station, Portland, Oregon.

Driver, B., Brown, P.J., Stankey, G.H. and Gregoire, T.G. (1987) The ROS planning system: evolution, basic concepts, and research needed. *Leisure Sciences* 9, 201–212.

Haas, G.E., Driver, B.L. *et al.* (1987) Wilderness management zoning. *Journal of Forestry* 85(1), 17–21.

Kuss, F.R., Graefe, A.R. and Vaske, J.J. (1990) *Visitor Impact Management: a Planning Framework*, two volumes. National Parks and Conservation Association, Washington, DC.

Leung, Y. and Marion, J.L. (2000) Recreation impacts and management in wilderness: a state-of-knowledge review. In: Cole, D.N., McCool, S., Borrie, W.T. and O'Loughlin, J. (eds) *Wilderness Science in a Time of Change Conference: Wilderness Ecosystems, Threats and Management*. USDA Forest Service, Rocky Mountain Research Station, Ogden, Utah, pp. 23–48.

Leung, Y., Marion, J.L. and Farrell, T.A. (2001) The role of recreation ecology in sustainable tourism and ecotourism. In: McCool, S.F. and Moisey, R.N. (eds) *Tourism, Recreation and Sustainability: Linking Culture and the Environment*. CAB International, Wallingford, UK, pp. 21–39.

Manidis Roberts Consultants (1997) *Developing a Tourism Optimization Management Model (TOMM): a Model to Monitor and Manage Tourism on Kangaroo Island, South Australia*. Manidis Roberts Consultants, Surry Hills, New South Wales.

Martin, S.R. (1990) A framework for monitoring experiential conditions in wilderness. In: Lime, D.W. (ed.) *Managing America's Enduring Wilderness Resource*. University of Minnesota Extension Service, St Paul, Minnesota, pp. 170–175.

McCool, S.F. and Cole, D.N. (eds) (1997) *Limits of Acceptable Change and Related Planning Processes: Progress and Future Directions*. USDA Forest Service Intermountain Research Station, Ogden, Utah.

McCool, S.F. and Lime, D.W. (2001) Tourism carrying capacity: tempting fantasy or useful reality? *Journal of Sustainable Tourism* 9, 372–388.

Merigliano, L.L. (1989) Indicators to monitor the wilderness recreation experience. In: Lime, D.W. (ed.) *Managing America's Enduring Wilderness Resource*. Minnesota Agricultural Experiment Station, University of Minnesota, Minneapolis, Minnesota, pp. 156–162.

Nilsen, P. and Taylor, G. (1998) A comparative analysis of protected area planning and management frameworks. In: McCool, S.F. and Cole, D.N. (eds) *Proceedings: Limits of Acceptable Change and Related Planning Processes: Progress and Future Directions*. USDA, Forest Service

Rocky Mountain Research Station, Ogden, Utah, pp. 49–57.

Stankey, G.H. and McCool, S.F. (1984) Carrying capacity in recreational settings: evolution, appraisal and application. *Leisure Sciences* 6(4), 453–473.

Stankey, G.H., Cole, D.N., Lucas, R.C., Petersen, M.E. and Frissell, S.S. (1985) *The Limits of Acceptable Change (LAC) System for Wilderness Planning*. USDA Forest Service Intermountain Research Station, Ogden, Utah.

US Department of the Interior (1997) *VERP. The Visitor Experience and Resource Protection (VERP) Framework: a Handbook for Planners and Managers*. National Park Service, Denver Service Center, Denver, Colorado.

Wagar, J.A. (1966) Quality in outdoor recreation. *Trends* 3(3), 9–12.

Further Reading

Gartner, W.C. (1996) *Tourism Development: Principles, Processes, and Policies*. Van Rostrand Reinhold, New York.

Munasinghe, M. and McNeely, J. (eds) (1994) *Protected Area Economics and Policy: Linking Conservation and Sustainable Development*. World Bank, Washington, DC.

CHAPTER 6
The Manager's Toolbox

Introduction

While the processes such as Limits of Acceptable Change (LAC) and Visitor Activity Management Planning (VAMP) discussed in Chapter 5 provide the overall framework for management of visitors, specific actions or tools need to be identified, implemented and monitored. Managers have available a wide variety of tools useful in achieving goals of enhancement, prevention and restoration. It is important when selecting specific tools and approaches to management that managers first identify the goals and values that the tools are designed to protect or enhance. Too often, much of management attention is focused on first identifying a tool and then determining how that particular tool will be used in a management situation. It is much better to start out by gaining agreement on goals and values before turning to the more specific management tools and actions. Planning processes such as LAC and VAMP provide managers and the affected public with the opportunity to identify those goals and values.

In this chapter, we provide a brief review of the tools available to manage tourism in national parks, emphasizing tools that are the least intrusive into a visitor's experience. We focus specifically on management of conflict because that issue dominates park planning and management, as noted in Chapter 4. Rules and regulations, which are often part of the manager's repertoire of available tools, are not directly covered. However, rules and regulations covering almost any aspect of human behaviour can be implemented if there is the legal basis for them and other tools have failed. Managers may also wish to manage visitors through the use of facilities, such as barriers, trails and bridges. These tools would again be subject to analysis to determine their appropriateness and effectiveness in any given situation.

Peterson and Lime (1979) provide a framework for considering what tools might be useful in any given situation (Fig. 6.1). Basically, management tools are oriented towards: (i) influencing visitor decision processes; (ii) controlling visitor behaviour; or (iii) mitigating the impacts of visitors. In general, because park visitation occurs during one's leisure time – when freedom of choice is particularly valued – managers should attempt to deal with provision of recreation opportunities and mitigation of impact problems caused by visitors first by influencing visitor decision processes. This involves communication of information about the park, its resources and values, recreational opportunities and appropriate behaviour to visitors, thus permitting them to make their own decisions about what they do, when, how and where. Such a strategy retains an internal locus of control over one's behaviour. If this particular approach is ineffective, then more intrusive techniques dealing directly with controlling, confining or regulating visitor behaviour – such as rules on where and when people can camp – will have to be implemented. Regulations and restrictions on visitor behaviour tend to be popular among managers, but often they may be difficult to enforce or ineffective, alienate visitors and negatively impact recreational experiences.

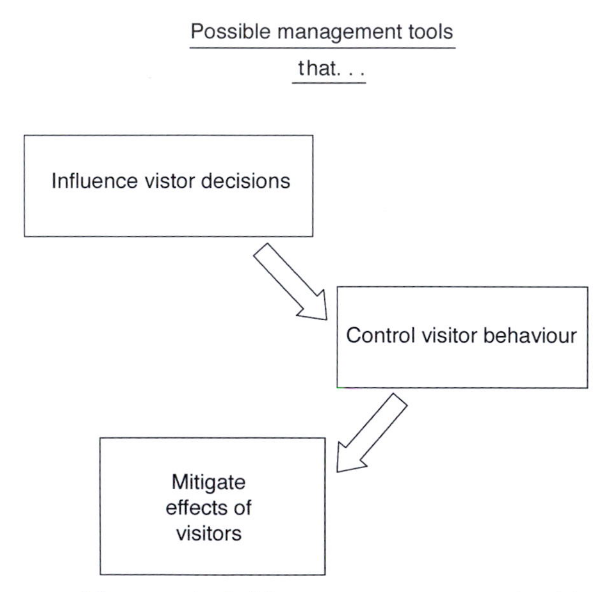

Fig. 6.1. Conceptual framework of visitor management strategies. Managers can choose from a combination of three different approaches to managing visitors (adapted from Peterson and Lime, 1979).

Finally, managers may wish to implement actions that directly mitigate impacts, such as hardening a site with blacktop or concrete, locating and designing facilities to minimize impacts, and confining visitors to specific pathways, roads and trails. Restoration of sites impacted may also be implemented. However, such actions may directly affect the character of the recreation opportunity provided, and thus must be analysed from the perspective not only of effectiveness but how experiences may be affected. For example, paving a rustic road with blacktop may control dust, but may also significantly affect the character of the recreation opportunity offered.

Within this context, managers may make decisions about visitors (using tools such as information, education and regulation), about facilities, such as trail location, design standards and construction techniques, and about the resource, such as site hardening techniques (actions that will help to resist the effects of recreational use). These techniques may be rather direct and intrusive, or they may be indirect and subtle. The techniques chosen will be influenced by objectives for the park or protected area, the scale of the management question (e.g. a few people or a lot of people are affected), the values involved, costs of development and operation, and perceived effectiveness. Often, more than one technique will be used in conjunction to increase their effectiveness.

Information and Education

Information and education are favoured techniques in national parks and similar protected areas. This comes from a belief that visitors – who in many cases grew up and currently live in urban areas – are unaware of appropriate behaviour, and from a belief that agencies should favour management techniques that are minimally intrusive into a park experience.

Information programmes can be directed towards a variety of issues, problems and opportunities. For example, agencies can develop information programmes aimed at reducing visitor-caused problems such as littering, excessive biophysical impacts, inappropriate campground behaviour, feeding wildlife and so on. Information may also be used to distribute visitors to different areas within a park that more closely match their preferences, enhance preparation for conditions encountered (e.g. signs in Sagarmatha National Park presenting information on how to deal with the potential for acute mountain sickness), distribute visitors temporally, and influence expectations of the opportunities that the park offers. Information may also be provided for basic directions and increasing awareness of facilities and programmes provided in the park.

The effectiveness of park information programmes in changing visitor behaviour has been studied extensively in the United States, and the results are often inconclusive or contradictory. Human behaviour is complex; in any given situation a variety of variables and influences will determine whether any management technique, including information, will be effective. These variables include (Vander Stoep and Roggenbuck, 1996):

1. *Precursor variables*: conditions within an individual such as attitudes, beliefs, values, knowledge of the area, its objectives and its rules.
2. *Context variables*: the social group with which the individual is recreating, demographics, expectations of social control, respect for management agency.
3. *External factors*: such as presence or opportunity to engage in alternative behaviours, presence of infrastructure that facilitates preferred behaviours, attitudes of rangers.

Thus, when considering information as a management technique, one must have an understanding of the values and beliefs of the primary audience, one must have carefully defined the problem or opportunity that is the subject of the information campaign, evaluated alternative media and timing of the information, and planned to measure the effectiveness of the information in changing behaviour. And as Vander Stoep and Roggenbuck observe, information may be most effective when a variety of media channels and other management actions are implemented in concert.

Education and information are techniques aimed at the visitor's decision-making process, not directly at mitigating impacts or directly controlling visitor behaviour. So education and information generally retain the locus of control within the individual but may not be as effective in reducing or mitigating impacts as more direct techniques. Education and information oriented towards impact management also assume that impacts result from ignorance or a lack of knowledge. Such an assumption may be questionable when impacts persist after information campaigns have been implemented. In some cases, visitors may be very well aware of appropriate behaviour but, because of the types of variables identified (e.g. peer pressure, situational circumstances) above, may not choose to engage in it.

Effective use of information and education requires that managers understand how visitors process information, the extent to which they accept the principles embodied in the persuasive message, situational factors and group influences on decisions to engage in the recommended behaviour. A wide variety of conceptual and theoretical approaches are available that serve as the foundation for persuasive message campaigns (Vander Stoep and Roggenbuck, 1996). These concepts should be consulted prior to development to ensure that the message will be effective in informing visitors and gaining their cooperation.

Interpretation is a technique closely related to information and education. Interpretation involves explaining natural or cultural phenomena in such a way that the visitor begins to understand the subject yet is stimulated to learn more. While interpretation may provide the 'facts' of a particular heritage value, the facts are included within a context that makes understanding such facts relevant to the visitor, outlines connections with other 'facts' and inspires the visitor to seek more knowledge and, in some cases, act on that knowledge. A major objective of such a perspective is to increase sensitivity of the heritage value involved so that visitors will support its protection. In the words of Freeman Tilden (1977), 'through interpretation comes understanding, through understanding comes appreciation, and through appreciation comes protection'.

Interpretation occurs through simple brochures and trails; films and slide presentations; guided walks and tours; and visitor centres. Such information-intensive activity suggests an opportunity for private sector involvement, particularly as guides and as businesses supplying displays and interpretive materials.

Regulation of Access

Limiting access to national parks and protected areas is one of the most controversial aspects of management. The controversy comprises a constellation of closely related issues that often embrace national philosophies regarding freedom, the role of government in controlling behaviour, cultural norms on access to public lands and how local communities relate to national parks. Access to publicly held lands varies considerably by culture and government. Various IUCN categories list the degree of access. We will discuss two major components: (i) regulating access under conditions of high demand with respect to supply; and (ii) prohibiting traditional uses of lands in newly established national parks.

Regulating access under conditions of high demand

When demand by tourists for certain recreation opportunities increases to the point where the values for which protected areas have been established are threatened or to the point where the capability of the agency to manage the numbers of tourists is stretched, managers may wish to adopt a use limit policy. Such policies generally regulate or limit the number of visitors that can enter an area by the day, week, season or month. Thus, a particular protected area may have a use limit of, for example, 200 people per day. Implementation of such a policy requires that the protected area agency has the administrative capacity to enforce the use limit. Such use limits are generally implemented through a required entry permit in combination with patrols to enforce the permit requirement.

Use limit policies tend to be controversial because they directly regulate and potentially restrict access to protected areas by visitors and tourists. Some visitors may find the requirement for a permit offensive. Usually, a required entry fee accompanies such permits. While many people support the notion of entry fees when the revenues from the fees support park management, others may object to them.

When the demand for an area rises above the amount permitted by the use limit policy, managers must employ some method of rationing the limited use opportunities that are available. For example, the Selway Wild and Scenic River in the State of Idaho provides spectacular white-water river floating (rafting, kayaking, boating and canoeing) opportunities, which are in high demand. The US Forest Service, which manages the river, limits the number of groups launching to one per day. However, demand for these opportunities is much higher and, therefore, a lottery to select the groups is held. Less than 3% of the groups that apply for a permit to float the Selway River are selected in the lottery.

Use limit policies involve four major decisions. The first decision establishes the overall use limit. This particular decision includes determining how many people will be allowed to enter the park per time period, what the appropriate measure of use will be (number of visitors, visitor-days, visitor-nights or some other appropriate measure) and how the use will be regulated: the administrative procedures. For example, a permit is required to float the Colorado River through Grand Canyon National Park in Arizona in the USA. Currently, the use limit is set at 169,950 user-days (a user-day is one person for 1 day on the river) per year.

The second major decision involves determining how much of the limited use opportunities will be allocated to the commercial and private sectors. Visitors without the skills or knowledge may wish to hire a guide or outfitter; others may feel they have these capabilities and visit the park without using these services. The primary question here is how much of the limited use permitted will be allocated to commercial services and how much will be allocated to people visiting on their own. In Grand Canyon National Park, the commercial sector is allowed 115,500 user-days per year and the private sector has been allocated 54,450 user-days. Once these limits are attained, no more use can occur on the river. However, these decisions are generally made early in the visiting season and are allocated in such a manner that use occurs continually throughout the season.

The third major decision is focused on the private sector. In this decision, managers determine how limited use opportunities will be rationed to individual groups or individuals (McLean and Johnson, 1997). For example, will rationing be based on a reservation approach, lotteries or waiting in line? The Selway River, as mentioned before, uses a lottery. The Grand Canyon uses a waiting line, which is now more than 10 years long.

In making such decisions, managers must be clear about not only the objectives of the rationing process but the administrative details as well.

Rationing may be designed to achieve various objectives. One objective may be to ration use by ensuring that each applicant has an equal chance of being selected. A lottery might be an example of a rationing system designed to achieve this goal. Another objective might be to consider need for access, for example, favouring those social groups – such as youth at risk – that might benefit most from a park experience. A third goal might be to achieve a socially efficient allocation. This would be done by using a pricing system to ration use. Finally, managers may choose equity as a goal. In this situation, those potential visitors who often bear the highest costs of a protected area, such as local residents, or who have higher inputs would be favoured. Each rationing technique has differing consequences for each of these particular goals and would affect different groups in different ways (Stankey, 1977). For example, a reservation system favours those who plan for trips far in advance. A waiting line or queue discriminates against people who are older (remember the 10-year wait for a Grand Canyon trip).

The fourth decision involves allotting use to individual firms in the commercial sector. There are currently 16 firms providing outfitting services for the Colorado River in Grand Canyon. Management must determine how much of the allocation for the commercial sector each firm receives. A major way this has been done has been to base allotments on historical use levels. As in private rationing, managers may want to consider what objectives the allotment system should be designed to achieve. For example, managers may want to ensure that a diversity of trip types or guide services are available.

One of the major issues that accompanies a rationing programme is the question of whether the use limit policy itself is more costly than other approaches to mitigating the impacts of tourist use. Use limit policies are administratively costly in terms of personnel, time and money. Not only are the allocation, rationing and allotment decisions difficult ones, but the administrative details in developing and implementing the requisite permits are often overwhelming. So, managers must make a decision as to whether they can accomplish the goals of the protected area more efficiently through a use limit policy or some other management technique. See Fig. 6.2.

Limiting traditional uses of national parks

In some situations, national parks and protected areas have been designated in places that contain indigenous peoples (villages and towns) and/or have areas used by them for grazing, crops, hunting or fuelwood. In many of these cases, residents have been removed from the park and their traditional agricultural and subsistence uses have been eliminated. The implicit proposition of such removals has been that permanent human habitation and use leads to an unacceptable level of biophysical

Fig. 6.2. Parks often have to balance visitor access demands with those of public safety and environmental sensitivity. These geysers in Yellowstone National Park are both attractive and dangerous to the visitors. The park provides relatively safe access by the use of interpretation messages and boardwalks across sensitive geological features. Paint Pots Geysers in Yellowstone National Park, USA. (Photographed by Paul F.J. Eagles.)

impact. While this may be true in some regions, in many areas, humans have continuously occupied landscapes for hundreds, if not thousands, of generations, and thus the landscape we see now is a result of the character of those long-standing interactions. Elizabeth Kemf (1993, p. 5) observes:

> many [protected areas] were also created without consultation with the communities that lived in or near them, whether they were by definition indigenous or other long-term residents. Ironically, it was these people who were for millennia the custodians of the earth, not always, but usually, caring for it so well that it had maintained its natural ecosystems in an unspoiled state. Frequently, when protected areas were established, indigenous and local residents were moved out, often to the detriment of the land itself.

Whether removing people from their homes and eliminating sources of food and fuel is the appropriate action for managing a park or protected area is something only those in the particular culture can judge. In fact, most parks and protected areas outside the Western Hemisphere currently have indigenous or local populations. For example, in England, where the original ten national parks were established to protect and preserve pastoral landscapes, farms, villages and even cities exist within park boundaries. In addition to the removal of local populations from within gazetted boundaries, the local knowledge, experience and expertise of community members concerning ecosystem characteristics and human interaction with it were often ignored or discounted in management decisions. These two actions, therefore, often resulted in considerable contention, conflict and in some cases have led to confrontation and violence.

Should subsistence and other pastoral uses be eliminated from national parks? Should indigenous villages and local communities be removed from within parks and protected areas? How should local, experiential knowledge be incorporated into park management decisions? These are significant and controversial decisions. When examining them, one must account for the customary and traditional uses of land, the objectives for each park and protected area, basic human rights of access, the impact of the activity on the values for which the park was established, the effect of removal of the activity on the individuals affected and other similar factors.

These questions are significant ones for many countries, local communities and indigenous peoples. The IUCN (World Conservation Union) Task Force on Local Communities and Protected Areas has identified a number of principles for dealing with these questions (see Table 6.1). Of course, these broad principles must be adapted to specific local conditions and situations, but do provide a framework for initiating discussion and conflict resolution processes.

Table 6.1. Guiding principles of the IUCN Task Force on Local Communities and Protected Areas.

1. Local communities are to be recognized as rightful, equal partners in the development and implementation of conservation strategies that affect their lands, waters and other resources, and in particular in the establishment and management of protected areas. This should apply to all IUCN categories of PAs, where local communities are present.
2. The livelihood security of local communities living within or around protected areas, and dependent upon the resource base within such areas, needs to be protected and enhanced while ensuring the ecological integrity of the area.
3. Since many local communities have a close link with natural resources, their traditional knowledge in conserving and sustainably managing their resources, and their own ways of valuing biodiversity, need to be respected and utilized in conservation measures.
4. Tenurial security of local communities over land, water and other resources, accompanied by appropriate responsibilities, is essential in creating and maintaining a stake in natural resource and biodiversity conservation.
5. Alternative and modified resource use practices need to be evolved, by or in association with local communities, to tackle unsustainable practices of resource use.
6. The principal benefits from conservation strategies and measures should go into further conservation measures and to local communities.
7. Forced displacement of communities that have traditional and customary rights to use of resources, in and around PAs, is unacceptable.

Managing Conflicts in National Parks and Protected Areas

Competition for the resources and recreational opportunities existing within a park or protected area often leads to conflict. The conflict may exist between the managing agency and visitors or communities within or adjacent to the park, among visitors, or between tourism development and environmental values within the park. Indeed, a large amount of agency effort actually goes into managing conflicts that continually develop. Such conflicts may be defined as interference with the pursuit of goals by one or more groups. Such interference is attributed to another group (Jacob and Schreyer, 1980).

The contentiousness and conflict that often accompanies management of parks is a natural outgrowth of competing visions of what parks and protected areas ought to be, of pluralistic societies that permit or encourage free expression of ideas, and of changing roles and functions of parks in any given society. The process used to resolve these conflicts is often as important to the parties to the conflict as the substantive outcome. In this section, we briefly review basic principles and concepts concerning conflict and its resolution.

Basic concepts of conflict management

The presence of a conflict can be viewed as the result of goal interference between two or more groups of individuals. In some cases, the conflict occurs between the park management agency and visitors or local residents, such as when park policy changes or the needs of the community are not adequately addressed in management plans. For example, when the management agency places prohibitions on long-standing customary subsistence uses of resources within the park, conflict will often result because local community residents can no longer secure necessary food or fuel. The prohibition interferes with the capability of residents to pursue a goal of subsistence. In other cases, the agency may wish to adopt new rules restricting some recreational uses, leading to conflict with visitors. In this situation, the impacts of the recreation activity may interfere with attainment of park goals, and park restrictions interfere with attainment of visitor goals.

Other conflict situations result when two or more visitor groups pursue competing activities, such as cross-country skiers and snowmobilers. In this case, the physical presence of each group may prevent the other from securing a satisfactory recreational experience.

Conflicts may be described along an asymmetrical–symmetrical continuum. Asymmetrical conflicts occur in situations where one group interferes with another, but the other group does not. Many snowmobiling–cross-country skiing conflicts originated as asymmetrical conflicts, for example. However, as a conflict goes unresolved, there is a tendency for the conflict to evolve into a symmetrical conflict, where each group perceives the other as interfering with the goals they are pursuing. The term symmetrical may not quite describe the situation correctly, because the basis for the conflict for each group may be somewhat different.

The basis for some conflicts is simply functional; the physical presence of one group may prevent another group from pursuing a recreational experience. In the case of snowmobiling and Nordic skiing, for example, snowmobile trails in the snow interfere with the trail needed by cross-country skiers. In other cases, the foundation for the conflict may be more value- or belief-based. To a cross-country skier, for example, not only do snowmobile trails interfere with the pursuit of their activity, but snowmobiles also produce noise and pollution that, while not having discernible physical effects on skiing, are nevertheless disliked.

The social science foundations for the presence of conflict among visitors or between visitors and agencies are many. They include differing values, beliefs and worldviews, motivations and expectations for specific recreational engagements, lifestyle differences and individual characteristics (e.g. ethnicity). For example, some visitors may enter a national park expecting solitude but be confronted with noisy and boisterous behaviour. In other cases, the lifestyles of some visitors may be viewed negatively by other groups.

Underlying many conflicts over use of national parks and protected areas are substantial areas of agreement and values held in common. When some goals or values are held in common, the conflict may be termed a 'mixed-motive' one. When no goals are in held in common, conflicts are 'zero-sum'. In zero-sum conflicts, the 'wins' of one group come at the expense of another group. For example, hardrock mineral mining in a national park directly conflicts with the goals of the park; this would be termed a zero-sum conflict because the presence of the park prohibits a mine, and the presence of the mine would destroy many of the values in the park.

However, mixed-motive conflicts are generally more frequently encountered in protected area management. For example, a conflict may exist between visitors who want to observe and photograph wildlife in an African national park and the managers who desire to protect wildlife from excessive disturbance. Both groups share basic values about wildlife and, at a fundamental level, both desire to maintain abundant levels of wildlife populations. This sharing of values allows development of management strategies that will benefit both wildlife and visitors.

Conflicts tend to escalate as they go unmanaged, resulting in polarization of user groups and the management agency. Escalation involves intensification of rhetoric as well as action: as a conflict escalates, parties to the conflict are more willing to engage in organization of advocacy groups, develop written petitions requesting or opposing agency decisions, or instigate non-violent demonstrations. As such conflicts escalate, it becomes increasingly difficult for the conflicting groups to communicate.

Some principles for resolving conflict

We noted earlier the inherent 'messiness' of park and protected area planning and management. Conflict is part of that messiness. We also stated that building consensus on proposed park actions is a necessary component of planning in any politicized situation. While such messiness and politicization may seem chaotic to many, there are, nevertheless, several principles for resolving conflicts in such situations. In this section, we describe four such principles that provide the foundation for successful conflict resolution.

Manage escalation We have noted the tendency for conflicts to escalate as they go unresolved. The increased polarization of conflicting groups leads to signficant barriers to resolution. Not only does polarization result in more difficult communication between groups, it often becomes more difficult for the park agency to communicate. For example, if the conflict is between two user groups, as the conflict escalates there is a tendency for the conflicting groups to view messages from the agency as supporting the other group's position, even if the message is relatively neutral.

Escalation will often lead to groups enlarging the domain of the conflict to include other groups and other arguments in order to secure additional support and sway resolution of the conflict in their favour. Thus, a local group may appeal to a regional or national group or in some cases attempt to obtain the support of international groups to put pressure on the agency. Such escalation makes resolution of essentially local conflicts more difficult and may lead to certain values or interests being overrun.

The park manager thus must manage the conflict to prevent unnecessary escalation. However, some escalation may be necessary so that the values and interests, policies or behaviours that are conflicting become more clearly identified and described. Managers would want to know who is affected and how. They would want to know the basis of the conflict: is it asymmetrical? Is it functional? Does it involve values or behaviours? Careful management of escalation may help to address these questions and clarify the situation.

Encourage definitions of the conflict that are mixed-motive in character Zero-sum conflicts are all but impossible to resolve by management. In fact, there is no motivation for the more powerful group in a conflict situation to negotiate a resolution. Zero-sum conflicts are resolved through sheer power politics. Such solutions in the long term breed dissatisfaction and lead to citizens questioning the legitimacy and credibility of government agencies and their decision-making processes.

As conflicts escalate, they are more likely to be defined as zero-sum conflicts, making equitable resolution more difficult. Thus, in order to resolve the conflict, managers must continually work to retain its mixed-motive character in order to maintain the opportunity for resolution strategies that are satisfying to all conflicting parties. This may mean first searching for the values, goals and interests that conflicting groups share. These may be very basic worldviews about human rights or democracy, or beliefs about the roles of parks and protected areas in society. Or they may be more situation-specific beliefs about a protected area.

Secondly, the manager may have to continually reinforce these shared values. This could involve presentations by the manager or even presentations by the conflicting groups themselves concerning their beliefs or values. The management action crafted to resolve these conflicts will be founded on these beliefs and values; making them explicit in the conflict management process emphasizes the common interests of conflicting groups.

Retain communication channels Mixed-motive conflicts can be effectively resolved through a variety of participative, collaborative and negotiation strategies. However, these strategies can only work if communication channels between conflicting groups are maintained. Such channels assume that conflicting groups are capable and willing to engage in face-to-face good faith discussions.

Maintaining communication channels is thus fundamental to the potential for resolving mixed-motive conflicts. There is a variety of communication channels, including the manager (if not a direct party to the conflict) serving not only as a conduit for messages but also as a conflict resolution facilitator.

Determine the interests at stake; avoid simplistic 'yes–no' solutions As conflicts escalate, stakeholders begin to develop and adopt 'positions' or statements of policy they want the agency to adopt. Often, the goals, values and interests these preferred actions reflect are only implicit. But arguing over actions before agreement on goals exacerbates conflict. For example, a park may wish to allow hunting of a large ungulate species in order to maintain some semblance of natural population dynamics because its predator has been locally eliminated. A local conservation group may oppose the proposal. If the conflict were simply defined as a matter of support or opposition to hunting, the fundamental interests of both sides may never be uncovered. It may be that both parties have interests in securing a healthy population of both animals and the vegetation they feed on. The conservation group may have an interest in humane treatment of the animals. The park agency may be attempting to meet some larger, socially prescribed mandates. Redefining the conflict in terms of each interest – human treatment, a healthy ecosystem, social goals – may allow both sides to develop creative solutions to the overpopulation problem, such as transplanting, sterilization, etc.

By uncovering interests, the conflict manager avoids the tendency to simplistically define a complex conflict and define groups as opposed to or favouring a particular management action. The focus now becomes securing, through a series of trade-offs, the interests of both groups rather than simple 'yes–no' responses to, for example, hunting. This avoids excessively reducing complex, dynamic conflicts to simply right or wrong solutions. The search for underlying interests first, then creative solutions (management actions) later becomes more difficult as conflicts escalate. As polarization occurs, groups are more likely to convey 'non-negotiable' positions and less likely to engage in joint factfinding, searching for mutual interests, and more unlikely to attempt to accommodate other values and interests.

Conflict resolution techniques

A variety of formal and informal techniques exists to resolve conflicts. The formal techniques (litigation, legislation and administrative rule making) are well known tools, but they are also tools that are not particularly useful in many conflict management situations. Moreover, many such processes, such as formalized administrative rule making (e.g. public hearings, formalized responses), often exacerbate local conflict situations by their adversarial character which tends to force conflicts into a zero-sum definition. In addition, such processes often remove control of the conflict from where it occurs to higher administrative levels. Therefore, in this situation, we will discuss collaborative or participatory planning, unassisted negotiation and assisted negotiation.

In collaborative or participatory planning, people representing affected interests or values (stakeholders) are deeply involved in planning processes. In participatory planning, stakeholders provide continuous input into planning and management decisions; such input may be guided by a formalized planning process – such as Limits of Acceptable Change – and stakeholders are asked to come to a consensus about potential futures and actions. Stakeholders include the affected public – preservation groups, commodity groups, local communities, concessionaires, agency decision makers and staff – as appropriate. These participatory processes are different from negotiation discussed below in the sense that they often involve comprehensive or general management plans for a park or protected area (rather than focusing on a specific issue, such as allocation of recreational use) and occur over a protected time period. Collaborative planning processes are often oriented towards not only preparing a plan but also establishing a framework for resolution of conflicts when they occur.

In these processes, it is important that all stakeholders have representation, have equal access to information, are treated respectfully, have a common definition of the problem or conflict in question, feel that the issue can be resolved with the help of the public, have identified specific issues and interests, and are willing to make necessary trade-offs. Table 6.2 summarizes many of the interpersonal processes involved in such participatory efforts.

At times, conflicts may escalate to the point where negotiation among stakeholders is needed. This negotiation may be unassisted or it may be assisted by mediators hired to help the stakeholders develop creative solutions so that various interests can be accommodated. Both types of negotiated situations will occur, however, only if all stakeholders

Table 6.2. A possible framework for resolving national park and protected area disputes (adapted from Moore, 1995).

Description of stage	Shared (social action)	Shared understanding developed	Managing agency influence	In-group relations	Out-group relations
Stage 1: Joint definition of problems	Joint realization and definition of problems	Joint definition of problems	High	Mistrust, blaming others for problems	Identity of group not yet established
Stage 2: Uncertainty about what to do	Uncertainty about how to deal with problems and best ways to protect interests	Mutual acceptance	Low	Extensive exchange of information, begin recognizing members' different roles and skills	—
Stage 3: Agreement on group procedures	Agree on ways of doing things	New way of doing business, setting the limits	High	Emerging sense of order associated with a developing, agreed way of doing things	—
Stage 4: Realization of interdependence	Realization of the need to work together to achieve anything	Development of group interests, realizing the group is the 'place to be'	Moderate (indirect)	Use of work groups beginning	Recognition of the group, transfer of information between group and outside world
Stage 5: Enthusiasm about collective possibilities	Enthusiasm about collectively influencing decision making	Development of group interests, realizing the group is the 'place to be'	Low	Emerging sense of joint purpose	Suspicion of cooption by outside world of group members
Stage 6: Commitment to working together	Commitment to working together to achieve group interests	Shared understanding about consensus	Low	Emerging collective identity	Members seek support for group from outside world
Stage 7: Consolidation of group	Members represent the group	Collective identity	Moderate	Shared understandings consolidate collective identity	Members take group decisions to outside world
Stage 8: Implementation of plan	Implementation of resolutions	—	High	—	Members support agency in implementing plan

involved agree that they will be 'better off', or more satisfied with a negotiated settlement than the current situation. Each party to a stakeholder will determine its 'best alternative to a negotiated agreement' or BATNA (Susskind and Cruikshank, 1987). If a group's BATNA is better than the outcome of negotiation, then the group will be unlikely to bargain. Part of the role of a manager is to understand the BATNAs for each group and determine how a negotiated agreement would provide groups with a better situation.

To a great extent, unassisted negotiations require a consensus among the affected stakeholders. A consensus may be termed 'grudging agreement', that is, some groups may be enthusiastic about the agreement and others may not like it, but for one reason or another (e.g. a poor BATNA) they are willing to go along with it. Building a consensus is not easy; there are many complete books of suggestions and conceptual foundations. It always involves developing clear expectations of what consensus would mean in a particular situation, joint factfinding, coming to a common definition of the issue and equal access to relevant information. Unassisted negotiations may lead to either written or informal agreements, but written agreements tend to make understandings explicit and reduce the likelihood of differences in interpretation of oral discussion.

Assisted negotiation often occurs in situations that have escalated and conflicting groups may have reached an impasse. None of the groups has another outlet (such as legislation or litigation) that would allow it to 'enforce' its position on the others, and its BATNA would not be very satisfactory. Assisted negotiation involves the agreement of the various groups that an outside, disinterested individual will be hired to mediate the conflict. Mediators have a variety of potential roles in any conflict situation, depending on the stage of the conflict and the needs of the conflicting stakeholders. For example, they might assist in joint factfinding by identifying qualified scientists and consultants, collecting and disseminating information and through iterative probing of the stakeholders about their perceptions. In some cases, the mediator may write a binding agreement for the stakeholders and in others may work privately with the various groups to enhance communications and understanding. The specific set of roles a mediator may play must be agreed on by the parties to the conflict.

In summary, conflict situations in national parks and protected areas are not only common, but they should be expected. A variety of approaches to resolving conflicts is readily available to protected area managers. Managers need to manage the conflict to ensure that it does not unnecessarily escalate with resulting polarization of stakeholders. Some of these conflicts can be reduced through participatory or collaborative planning processes; many others will develop out of changing social needs and values, and require some type of negotiation, either unassisted or with the help of a mediator.

Conclusion

Managing park visitors is not inherently conflictual, but should be conducted on a foundation of scientifically acceptable principles and socially appropriate values. Understanding these principles and values is a prerequisite to implementing any management action. Often, managers begin with a problem, issue or question and attempt to implement management actions to solve it.

But implementing such actions outside a decision-making process often does not solve the problem, but simply displaces it elsewhere in time or space. Structuring visitor management through a process such as LAC or VAMP makes decisions explicit, transparent and increases the potential for visitors and other stakeholders to understand and support them. Such explicitness also allows the value judgements inherent in park management to be displayed and debated, necessary deliberative processes. An important component of processes such as LAC is their foundation in the fundamental goals and values contained in the park. By first emphasizing, discussing and agreeing on these, stakeholders identify interests they hold in common; for interests in conflict, the common ground forms the foundation for resolving conflicts (Fig. 6.3).

Essential to any management process is the monitoring that is part of it. Monitoring is not an option; it would be unethical to implement a management action without seeking to understand its consequences and

Fig. 6.3. Planning for the future of parks, and solving the challenges confronting them, is a political process. Capital funds for the construction of major facilities are popular with politicians because of the public exposure possibilities. The political payoff was obvious at the opening of the Algonquin Provincial Park Visitor Centre in 1993. Operating funds for the day-to-day park maintenance provide little payback for politicians and are therefore much more difficult to raise from governments and private donors. Opening of the Visitor Centre, Algonquin Provincial Park, Ontario, Canada. (Photographed by Paul F.J. Eagles.)

effectiveness. Monitoring provides the necessary information to evaluate how well a plan is meeting its goals and whether there is a need to change actions.

Managers have a variety of tools available to deploy in order to achieve objectives. These include information, education, interpretation, zoning, use limits, restrictions on visitor behaviour and so on. As they deploy these actions – or if they fail to act – conflicts will arise. While managers have a variety of conflict resolution approaches available, early and intimate involvement of stakeholders in planning not only will allow many of these issues to be successfully resolved, but such participation establishes a foundation and set of relationships useful for resolving conflicts in the future.

Parks, as we have noted in several places, are political manifestations of a society's interest in protecting its natural and cultural heritage. To a great extent then, planning for the future of these parks and solving the challenges confronting them is a political process as well. A park manager, then, is as much a professional as a politician; as much a facilitator of people's interest as a naturalist; as much a mediator as a ranger.

References

Jacob, F.R. and Schreyer, R. (1980) Conflict in outdoor recreation: a theoretical perspective. *Journal of Leisure Research* 12, 368–380.

Kemf, E. (1993) In search of a home: people living in or near protected areas. In: Kemf, E. (ed.) *The Law of the Mother: Protecting Indigenous Peoples in Protected Areas*. Sierra Club Books, San Francisco, California, pp. 5–11.

McLean, D.J. and Johnson, R.C.A. (1997) Techniques for rationing public recreation services. *Journal of Park and Recreation Administration* 15(3), 76–92.

Moore, S.A. (1995) The role of trust in social networks: formation, function, and fragility. In: Saunders, D.A., Craig, J. and Mattiske, E.M. (eds) *Nature Conservation 4: the Role of Networks*. Beatty and Sons, Surrey, New South Wales, pp. 148–154.

Peterson, G.L. and Lime, D.W. (1979) People and their behavior: a challenge for recreation management. *Journal of Forestry* 77, 343–346.

Stankey, G.H. (1977) Rationing river recreation use. In: Lime, D.W. and Fasick, C. (eds) *Proceedings, River Recreation Management and Research Symposium*. North Central Forest Experiment Station, USDA Forest Service. General Technical Report NC-28, St Paul, Minnesota, pp. 397–401.

Susskind, L. and Cruikshank, J. (1987) *Breaking the Impasse: Consensual Approached to Resolving Public Disputes*. Basic Books, New York.

Tilden, F. (1977) *Interpreting Our Heritage: Principles and Practices for Visitor Services in Parks, Museums and Historic Places*, 3rd edn. University of North Carolina Press, Chapel Hill, North Carolina.

Vander Stoep, G.A. and Roggenbuck, J.W. (1996) Is your park being 'loved to death?': using communications and other indirect techniques to battle the park 'love bug'. In: Lime, D.W. (ed.) *Congestion and Crowding in the National Park System*. Minnesota Agricultural Experiment Station, St Paul, Minnesota, pp. 85–132.

CHAPTER 7
Monitoring of Tourism in National Parks and Protected Areas

Assessing the Impact of Tourism

The passage of the National Environmental Policy Act of the USA in 1969 (CEQ, 2001) launched the idea that the environmental impacts of government actions and the alternatives to those actions should be studied before decisions are taken. The concept of environmental impact assessment, as it became known, quickly spread globally (IAIA, 2001).

Most park agencies are now required by law or policy to study the environmental impacts of their policies and procedures. In most cases the environmental impact studies involve impacts on biological and ecological features of parks. In many cases these studies also look at the impact on social, economic and cultural features as well. There are often several levels of study. The broadest level involves a full impact study of all park activities, usually associated with the development of a park management plan. Often more limited studies are also required when individual developments are proposed, such as a new road, a building or a trail. Since park tourism is an important and widespread activity in many parks, tourism impact assessment is a valuable tool in all stages of planning and management (Fig. 7.1).

Impact assessment utilizes a very wide range of tools to measure existing conditions and to predict impacts on those conditions from proposed actions. It is a highly specialized field.

A key aspect of impact assessment planning is the evaluation of the importance of an impact. For example, a biologist may predict that a certain percentage of a large mammal population may change migration patterns because of a proposed trail in a mountain pass. Is this impact significant? What level of impact is required before the trail is rerouted or abandoned? Over time it has become obvious that the determination of value is the key aspect to all impact assessment. Chapter 1 outlined the range of approaches used to determine value in parks. Similar approaches can be used in impact assessment.

Chapter 5 outlined the Limits of Acceptable Change approach to the determination of the allowable levels of impact and recreation use. The key aspect of this approach is the process used to arrive at decisions. Similarly, the key aspect of impact assessment is the process used to arrive at value determination and, therefore, the allowable levels of development and impact.

Central to all impact assessment is the prediction of impacts, given specific environments and proposed actions. Such predictions are

Fig. 7.1. Monitoring of mortality from vehicle collisions revealed severe impact on the large mammal populations of the Bow Valley in Banff National Park. Aggressive and expensive actions were taken to reduce this impact. The highway was fenced and large underpasses constructed to allow wildlife movement under the road. When it was discovered that some predators, most specifically wolves, would not use the tunnels, a large overpass was constructed. Wildlife underpass and roadside fence in Banff National Park, Canada. (Photographed by Paul F.J. Eagles.)

much more accurate when information is available on the impact of past developments. For instance, given our example of mammal migration, when the biologist has scientific study results available that measured the impacts that occurred when certain types of development took place in other migration corridors, the accuracy of predictions of the impact of future development is improved. Therefore, central to all park planning, park management and impact assessment is the availability of accurate monitoring data from existing parks. Therefore, a monitoring programme is very important in any park tourism management regime.

Monitoring the Impact of Tourism

People visit parks and protected areas for many reasons. In order to have their goals fulfilled, people require facilities, services and programmes. Transport facilities are constructed, safety services are developed, information programmes are provided, food is purchased, photos are taken, accommodation is used. A large number of people, environments and communities are changed by the flow of people, money and vehicles associated with park tourism. Clearly, park visitation has social, cultural, economic and environmental impacts. Some impacts are inevitable; however, they can be planned for and managed.

For any planning and management to be suitable, it is important that there is a solid understanding of the impacts of a phenomenon. A prediction is only as good as the information base, the predictive models and the skill of the planner.

Monitoring involves a programme of impact measurement. Typically, park managers develop a programme to measure the impacts of specific programmes and activities (Fig. 7.2). These can range from relatively simple measurement of visitor volume, through more sophisticated measurement of recreation use on a sensitive environment, to a full economic impact calculation of tourism.

There is an abundant body of literature on the impacts of park tourism generally and the impacts revealed at specific sites. For any particular park, it is important that tourism impact monitoring takes place. Given the significance of good monitoring data for the understanding of impacts and trends, it is surprising how infrequently parks undertake solid tourism monitoring programmes. Such programmes are the exception rather than the rule.

The simplest and most important factor in tourism monitoring is the measurement of key *visit attributes*, such as the number of people entering the park and their length of stay. Everyone, from the park visitor to the park manager, likes to know the level of public use of a park. This use level is critically important in all impact prediction models, such as the development of an economic model or the understanding of the forces causing wear and tear on a nature trail. It is often valuable to measure

Fig. 7.2. Monitoring involves a programme of impact measurement. A lack of impact standards results in a lack of monitoring. Over time the tourism vehicles eroded away this sand dune in Fraser Island National Park. A lack of erosion impact standards and the resultant lack of impact monitoring allowed much erosion to take place before corrective action was ultimately taken. Road erosion in Fraser Island National Park, Australia. (Photographed by Paul F.J. Eagles.)

additional visit use attributes, such as type of transport used, the level of use of commercial services, temporal use distribution and spatial use distribution.

For most visitor management, it is important to understand various aspects of *visitor attributes*. Commonly measured attributes include socio-demographic characteristics, level of experience with a site or an activity, knowledge of environmental conditions and park regulations, and various attitudes towards management practices, park services and environmental conditions.

The *effects of the visit* need to be known. There is potentially a full range of ecological, economic, cultural and social impacts of park use. Individual studies of these impacts are done by highly trained specialists in specific fields. The study results guide managerial action in directing visit management policy.

Models for Public Use Measurement

Parks and protected areas are distinctive and attract significant public interest. Public interest leads to a stream of visitors who invest large amounts of money, time and effort to experience these areas in person. Many factors influence the experience of visitors, including: the conditions in the resource itself, the logistical support available in the park or in the local area, and the attitudes of people contacted, including the park staff and other visitors. When visitors return home, many of them become articulate and important voices favouring legislative support for existing parks and the creation of new sites.

Monitoring public use is a fundamental responsibility for managers. The resulting numbers are critical indicators of the natural, social and economic functions performed by parks and their caretakers.

The very factors that make parks and protected areas unique and exciting also make the necessary measurements of use difficult. Where great distances are involved, staff time is consumed by the logistical demands of transport to areas where measurement takes place. Park boundaries may enclose villages and residences as well as roads and trails, which necessitates the use of calculations to adjust measurements so they conform to the basic rules of reporting. Local residents near the park and protected area may visit in high numbers but carry out activities distinctly different from foreign tourists. Such complications of the human ecology surrounding and occupying the park require careful visitor studies as well as basic volume counts.

For the comparison of public use measurements, over time and between sites, it is necessary to adopt a standard set of definitions of the terms and concepts involved. The World Commission on Protected Areas has developed standard definitions and approaches for public use measurement in parks and protected areas (Hornback and Eagles, 1999). In order to provide an understanding of the concepts involved in public use measurement, it is necessary to present some of these definitions.

The World Tourism Organization developed a definition of a tourist.

> *Tourist*: a person travelling to and staying in a place outside their usual environment for not more than one consecutive year for leisure, business and other purposes.

This definition involves two elements: travel of a certain distance from home and a length of stay. For most parks, a percentage of the visitors are tourists, the rest are considered local residents. The definition of the area 'outside their usual environment' varies between countries. For example, Canada typically uses an 80-km distance (50 miles), while the USA typically uses 160 km (100 miles). Sometimes park agencies do not follow their national tourism statistics definitions and develop their own. For example, the Ontario Provincial Park agency uses 40 km, making their data incompatible with national tourism statistics. Clearly, standardization of definitions, such as the length of travel for a person to be considered a tourist, is essential for comparison between parks, park systems and with other travel statistics.

> *Entrant*: a person going on to lands and waters of a park or protected area for any purpose.

All people who enter a park have an impact. Therefore, it is important to know the total number of people entering a park. However, not all people entering a park are doing so for recreational or cultural purposes. Some may be simply travelling through on a major road. Some may be there to repair the road. Others may be bringing a shipment of food to the park's restaurant. A visitor is a person who visits the park for the purposes mandated for the area, typically outdoor recreation or cultural appreciation. A visitor is not paid to be in the park and does not live permanently in the park. Therefore, it is important to differentiate between park entrants and park visitors.

Visitor: a person who visits the lands and waters of a park or protected area for purposes mandated for the area.

A person who visits a park for part of one day has a different impact from someone who visits for 7 days. Therefore, it is important to keep track of the length of stay of park visitors. The combination of the statistics of the number of visitors and the length of stay leads to a new statistic, the number of visitor days. One person staying in a park for 1 day provides 1 visitor day of use. One person staying in a park for 7 days provides 7 visitor days of use. A park occupied by four people for 4 days gives 16 visitor days of use.

It is important to consider the number of hours constituting one day of visitation. Most parks accept a visit for any length of time during a 24-h period as constituting a day, as long as the person does not stay overnight. The length of stay is often ignored because the park does not have the ability to measure both the entrance time and the exit time of each visit. However, those parks with the ability to measure the time of entrance and the time of exit of a visitor can calculate the length of stay in hours. In this case a visitor day may be defined as 6 h of visit or 12 h of visit, two typical definitions. There can be confusion in the addition of visitor-day data from one park to the visitor day data from another park, when the number of hours of the visit is not used in the calculation of the visitor-day statistic. For example, the average length of stay for natural environment parks' day use is from 6 to 8 h. Typically, the average length of stay for historic parks is shorter: 2 to 4 h. Therefore, the total number of hours of recreation per visit is different between the two types of parks, but the overall visitor-day data may be shown as quite similar. In this example, if the parks accepted 6 h of visitation per person as constituting 1 visitor day, then the natural environment park would show 1 visitor day per visit and the historic park 0.5 visitor days per visit.

Visitor days: the total number of days that visitors stay in the park for a purpose mandated for the area.

It is common for people to stay overnight in a park, in a campground or a lodge. Many parks keep special statistics on the number of visitor nights occurring in a park. These are sometimes referred to as camper nights, bed nights or lodge nights.

> *Visitor nights*: the total number of persons staying overnight in a park or protected area, for a purpose mandated for the area, × the number of nights each stays.

Some park agencies see a necessity to combine visitor day data with visitor night data, to provide an overall use level. Others keep the data separate.

For comparison between parks and for important calculations, such as economic impact estimation, it is critical that parks maintain an accurate tabulation of total visitor days and visitor nights of use. Many other definitions are required for the full development of a public use measurement system. These definitions can be found in Hornback and Eagles (1999).

There are many techniques available for the actual measurement of visitor use. A common approach is that of the purchase of a permit. This typically involves people going through a control point where a fee is charged and a permit for entry is provided. This system is highly accurate but expensive to operate. An allied system is that of registration. This involves people voluntarily filling out a form. This is sometimes done before the visit and is called preregistration. It is sometimes necessary to undertake mechanical counts of visitors or their vehicles. This can be done with counters on a road that record all passing vehicles or with counters on a trail that record the passage of all hikers. Visual observation can be used to record information from a distance on visitors. This method can be used to check the accuracy of other methods, such as mechanical counts or voluntary registration on a trail. Indirect estimation involves the prediction of a characteristic from some other variable. For example, one could estimate the number of people entering a visitor centre by looking at the daily sales of the bookstore. Over time, data could be collected that provides evidence of the relationship between visitor numbers and sales, and then numbers can be estimated from sales. Remote sensing methods are sometimes used for remote areas or high-density areas. For example, satellite or aerial photographs can be used to count the number of people in canoes on a lake or on a beach. One of the most common techniques is that of a visitor survey. This is the method of collection of advanced levels of information about visitors and their characteristics. Many of these techniques are combined in a park's public use measurement system.

The application of a public use measurement system requires a full understanding of all aspects of survey design, sample design, statistics and error estimation. It is critical that a full understanding of limitations of the system be utilized to design an appropriate approach and in the reporting of the data. Watson *et al.* (2000) provide a summary of the methods available for the design and application of a public use measurement system within a wilderness type of park.

Using Tourism Statistics in Decision Making

The size of a park's public use measurement programme should be in proportion to the needs of park managers to provide for general management, natural resource protection, maintenance operations and visitor services and protection. Examples of the role of visitation data in these categories are outlined below.

General management

All managers need quantitative data on how visitation affects the park or protected area, and qualitative data on how the park or protected area affects the visitor. Moreover, managers need to know these conditions internally as well as immediately outside the park. Current public use volumes, overnight stays by type, rate of change, comparisons to other local areas and related data are of value to local residents, officials and businesses, as well as to other agencies and to general government functions. All people who plan for park visitation must have good data.

Current volume in a park, possibly reported on a weekly basis, gives an idea of the popularity of various activities and services. These data are usually collected at the point of delivery, such as roadside display use, interpretive programme attendance, concessionaire-provided guide, boat or livery service demand. Measurements from zones of use and avenues of access can indicate optimal fee collection locations as well as locations for allocation of staff resources and possible staging of facilities and services.

Visitation data assist in the computation of statistical reports on sanitation, public health, accident, fire suppression, criminal acts, search and rescue missions, etc. These data have expensive consequences for management. They direct actions that can involve considerable numbers of people and resources. They are especially important during the government budgetary process.

Visitation data are convertible into economic consequences (tourism sales, jobs, taxes) and help to evaluate the value of the park and park resources in common terms with other activities (agricultural, mining, etc.). Visitation data provide useful insights for people who might otherwise not care or understand the influence of the park on them and their economic lives.

When unusual events occur and an idea of impacts on visitation is needed to deal with unexpected consequences or emergencies, accurate and comparable historical data are suddenly crucial. In other words, current visitation data often become critically important at some later point in time.

Natural resource protection

The protection and management of natural resources is an important activity of most parks. Visitor use has a direct and immediate impact on a park or protected area's natural environment. Of special interest is any subsistence-based resource use, such as may occur by local or aboriginal residents.

Knowledge of public use activity, location and volume is needed to evaluate and preserve viable natural ecosystems, including endangered and threatened plant and animal species. Such data are very important for wildlife habitat management, prevention of human–wildlife interaction problems, protection of range, migration patterns, resting and nesting sites and the maintenance of vegetative cover, soil surfaces and water quality within desirable limits.

Knowledge of planned public use activity, location and volume is necessary to evaluate, protect, preserve and maintain cultural sanctuaries, archaeological sites and historical structures. Attention paid to visitor volume leads to better awareness of general visitor behaviour, from cutting firewood to the management of sanitary needs. As a result, preventive measures can be taken, ranging from signs to facility construction.

Some visitors explore away from the designated park roads and trails. As they leave traces for others to notice, new trails emerge. Resource managers must know total use, use at planned sites and unplanned visitation around critical sites. They must watch for, detect and deal with changing internal use patterns. Fractional changes can lead to rapid impacts on fragile resources, sensitive wildlife, migratory zones or delicate habitats and require immediate deterrent measures. Visitation rates, rates at sensitive areas, known peak loads at nearby areas and records of resource wear at certain volumes must be regularly analysed to protect and maintain the natural resources within the limits of acceptable change.

Maintenance operations

While maintenance operations can be performed after damage is evident, preventive maintenance operations require knowledge of visitor use levels. Public use volume and short-term forecasts of use are needed to order supplies and maintain minimal inventory of consumables (soap, toilet paper, paint, petrol, etc.). Public use volume data at specific service areas (campgrounds, picnic areas, roads and trails, parking and staging areas, etc.) are needed for daily repair and maintenance as well as replacement budgeting and scheduling.

Current, peak and seasonal volume at each major site in a park can be examined for changes in utilization rates relative to capacity, age and useful facility life cycle, changing rates of routine maintenance, and replacement cost programming or associated consequences of prolonged use. Records of public use volume can be combined with visual evidence of excessive wear to evaluate needs for maintenance. Once the associated costs are estimated, a decision can be made whether to continue preventive maintenance or to redesign and reconstruct the facility.

Park maintenance staff levels should be assigned on the basis of internal concentrations of visitor activity for clean-up and hasty repair. The public are very willing to complain if they see insufficient maintenance or observe too many people at a site. Therefore, assignments need to be made in sufficient strength, with the right equipment and at the right time.

Visitor use data are critical for road planning and design. Of special interest are data associated with local traffic that may use park facilities for local trips only. These people are typically not considered park visitors, but they can place considerable pressure on park facilities.

Visitor services and protection

Visitor services and visitor protection are two key elements of all park visitor management. The provision of visitor services is dependent on the needs and numbers of visitors, the availability of funds and the resource protection concerns (e.g. site hardening, pollution abatement, etc.). Establishing and maintaining public use safety and sanitation standards must be conducted in ratio to actual volume (water quality sampling and treatment, waste removal, etc.).

Daily operational service level standards (especially instructional directives to visitors and changes in monitoring and patrol functions) need to be made according to use volume as it relates to season. Many parks have dramatic seasonal differences in use levels. Current temperature conditions (extreme heat, cold or rapid weather change), resource conditions (fire hazard warning and restrictions) and wildlife control activities (preventive measures associated with dangerous animals or unusual disease threats) all affect service levels.

Public use monitoring activities put park staff in a position to detect, control and correct restricted or illegal activities (poaching, removal of artefacts, destruction of plant materials, etc.). The visibility of park staff during various public use monitoring activities has a secondary benefit of deterring vandalism.

The park may have legal liability exposure during certain times of use, such as periods of very high use or times of participation in dangerous activities. The volume and timing of such activity needs to be known, communicated to appropriate people and contingency plans made.

Public-use monitoring activities make staff accessible to the public to address needs or provide impromptu environmental education. Monitoring activities help to ensure that adequate volumes of materials required for public distribution can be ordered and sufficient inventory maintained. The provision of interpretive programmes and information services is frequently tied to anticipated visitor numbers based on records of previous volumes measured.

Establishing a Statistical Database of Visitation

All public use reporting involves the investment of effort for: (i) establishing the goals of the system; (ii) adopting the appropriate methods of data collection; (iii) collecting data; (iv) summarizing data; (v) analysis of data; and (vi) presentation and interpretation of data for management action. The exact size of the investment made by the managers in monitoring public use depends on the needs of management. More sophisticated needs involve more sophisticated data collection and analysis techniques.

The World Commission on Protected Areas suggests that there are five progressive levels available for a public use data management programme. Each higher level results in greater accuracy and detail of public use data and a corresponding increase in cost due to staff time, hardware and funding. The *Initial*, Level I, of a public use reporting programme is the simplest system. A Level I system involves park staff keeping track of observed use as they go about their normal duties such as park policing or visitor information duty. This system is inexpensive and provides a rough idea of use levels, but has low accuracy. This system provides almost no information on the visitors beyond roughly estimated use level data.

The *Basic*, Level II, system involves park staff developing procedures to provide partially accurate levels of counting. Typically, procedures are put into place to collect data systematically, with special consideration given to reducing double counting and to count those visitors who were missed in the Level I approach. The Level II system relies on park staff collecting data in addition to their other duties. A Level II system has more reliable use level data than a Level I programme, but nothing much is known about the visitor.

The *Intermediate*, Level III, involves systematic data collection of park visitation, with the introduction of sophisticated survey methodologies to provide sample data on key issues such as length of stay, home location and sociodemographic characteristics. At Level III some park staff members have sufficient training, resources and encouragement to supervise and undertake more accurate measurement of use levels, and some information on visitor characteristics. Most highly used parks have visitor data collection systems approximating the Level III system of measurement.

With a *Developed*, Level IV, programme the park has sufficient financial resources and staff to administrate and support a public use reporting programme at a level of accuracy that will serve the needs of all operational departments including planning and budget. Such data are also sufficiently accurate to be used in political arenas beyond the park, say with the local political leaders.

Three major actions raise the park or protected area from the *Intermediate* level to the *Developed* level.

- The visitor survey is replicated frequently, at least every year and possibly every season. Seasonal adjustments to the automated instrument system generate public use data and estimates for all areas of park use, including entries and exclusions, such as data from non-visitors.
- The administration of the overall system is supervised by one staff person with specialized skills who controls all aspects of the programme (instrumentation, operation, compilation and reporting) including consultancy to other operational departments for their public use and visitor data needs. Data accuracy and errors are the responsibility of this person.
- Instrumentation is installed, where appropriate and cost effective, based on an analysis of visitors and their traffic through the park. At a minimum, this means an optical sensor to count visitor centre traffic and entrance station traffic counters. Errors from hand counters and staff trying to do two jobs at one time are eliminated.

The Level IV system is the desirable system for most parks with higher levels of tourism use. The Level IV system has the depth and accuracy that stimulates confidence in its use, at least by those with sufficient knowledge to understand the limitations involved in data collection using Levels I, II or III.

Parks with high volumes of use, extensive development or planned development or critical and threatened resource conditions are best served by the precision, depth and usefulness that is associated with an *Advanced*, Level V, programme. Attributes of Level V include:

- trained staff dedicated to the programme;
- the use of sophisticated, remote counting equipment, such as road counters, trail use counters, remote sensing;
- computerized data collection on all those who register for use, such as campers, wilderness trail users or information seekers;
- the use of computers, enhanced graphic and statistical presentations of data;
- online access to data around the park and at remote offices;
- real-time understanding of current use levels; and
- supplementary detail for all park operating departments.

At Level V, visitor surveys are conducted frequently. The surveys contain policy and issue-specific items, as well as statistical and research-oriented questions. Studies might include data about transport, evaluation, activity or perception. At this stage the applications of data become especially important, not just to the park itself but also to the entire system of parks it represents. Data on the economic impact of park use and the kinds of specific expenditures people make are usually highly valued by local businesses and political leaders. A Level V system provides park use data that can confidently be used for regional economic evaluation.

Watson *et al.* (2000) provide an excellent review of many of the most important issues involved in establishing and operating a public use measurement system. The full application of this report's recommendations leads to a Level V system.

All those who use park use data should be cognizant of the quality of the data. What is the level of accuracy of the data? What is the error rate? Are the data an underestimate or an overestimate? Is there any reason for the data to be purposely misleading?

Visit Attributes

Each visit statistic has attributes. The most common visit attribute reported in parks is that of total visitation. Most parks, but not all, provide yearly figures of total visitation. For example, the National Park Service of the USA reported that in 1999 total visitation was 287,130,879 (NPS, 2001). It appears that this figure is composed of visitor days of activity. The complexity of collecting such information from 380 areas covering more than 33.7 Mha is very high. As a result, it is critical that the National Park Service uses standardized approaches and makes the methodologies known to all who wish to use such valuable information.

One commonly measured visit attribute is group size. This is important for facility design and site operations. For example, if a park is measuring visits by campsite registration, it is necessary to know group size so that the total number of campers can be calculated by multiplying the number of campsite registrations by the number of people in each group. Table 7.1 shows the information on group size of Ontario campers as obtained by a visitor survey in 1996. The minimum group size was one person, and only 4% of all campers camped alone. The maximum group size was seven, and 278 groups were this large. The most frequent group sizes were two people and four people. The mean camping group size was 3.30 and the standard deviation was 1.43. Such data are very important for campground managers. These data are useful in many aspects of

Table 7.1. Group size of Ontario Provincial Park campers.

Number of people per group	Frequency	Percent
1	471	4.0
2	4417	37.3
3	1561	13.2
4	3025	25.5
5	1461	12.3
6	630	5.3
7	278	2.3

management, such as estimation of facilities needs (e.g. number of showers required) and supply needs (e.g. the amount of toilet paper that must be ordered).

Visitor Attributes

There are many visitor attributes that are important for management. These include such information as visitor age, education level, home location and many other important factors. Figure 7.3 shows the age distribution of campers at one park, Killarney Provincial Park in Ontario. For comparison purposes, the age distribution of all park campers in the 1996 visitor survey from 44 parks is included. In addition, the age distribution of the Ontario population is included. Figure 7.3 shows several findings that are important for management.

The Killarney campers are more likely to be 25 to 44 years of age than are campers in other parks and the Ontario population. Conversely, Killarney attracts lower numbers of children and seniors than do other parks. These findings are important for many aspects of park management, including the design and delivery of the interpretive programmes offered in this popular park.

Figure 7.3 also shows a drop in camping with age. Ontario parks capture a much smaller proportion of the Ontario population over the age of 45. This capture rate declines dramatically with age. Both Killarney Park and provincial parks in general have a severe problem in attracting campers over the age of 45. This is a very important age group with higher levels of disposable income and more leisure time owing to lowering levels of family responsibility. This is the group that tourism businesses all over the world try hardest to attract, and the park's agency is doing

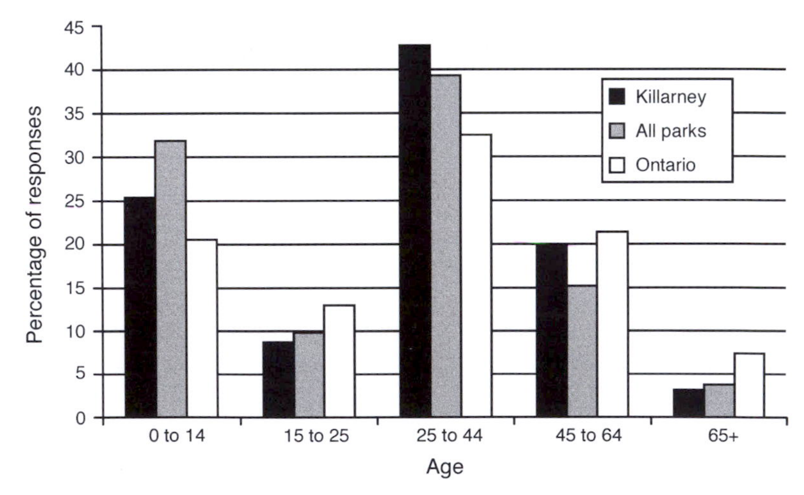

Fig. 7.3. Killarney Provincial Park camper age distribution.

poorly with this market segment. This analysis is a simple example of the use of demographic data, taken from a camper survey, for comparison with general population trends.

The presentation of visitor attribute data can be done in table or graph format. It is important that the data are presented for the personal interpretation of the readers. However, it is also important that the tourism management specialists who do the analysis and prepare the findings explain key findings in the data. Those who are not trained in the field of data analysis can benefit from a professional data interpretation.

Surveys of Tourists' Satisfaction: Importance and Performance Measures

We all have a choice on how to spend our leisure time and money: we could go to one park or to another; we could buy this product or that service; we can recommend a product to someone or we can caution against such a product. People visit a site, buy a product or recommend a service when they expect a level of satisfaction greater than something comparable.

Many parks care little about the satisfaction of their visitors. These parks typically provide take-it-or-leave-it levels of service. In other words, if people come and use the park, fine, if they do not, this is fine too. Such parks have little concern about their public profile or their level of political support. How can such a situation arise? This typically happens when park budgets are provided by government, irrespective of public use levels, visitor satisfaction levels or tourism impacts. This may occur when the goal of the park is only indirectly tied to visitation, as may occur when the park is justified solely on resource protection grounds. Or it may occur when the park staff members are entirely trained in resource management, with little understanding or sympathy for visitor use. This may also occur when the park budgets are so low that few if any park staff are available to measure visitor satisfaction and to take action on these measurements.

However, the societal and political environment for parks is such that parks that ignore visitor satisfaction may find themselves with little budget or, worse, are transferred into some other land use that has a higher societal value. Increasingly, many governments are requiring parks to earn from tourism all of the income that is necessary to manage tourism. In this situation, parks must function like any business. They must develop a market that is sufficient to support the parks' activities.

Highly satisfied visitors are more willing to pay for their services, give better recommendations to others and are more likely to return. Satisfied visitors are more likely to donate money to conservation and support political initiatives for conservation. Therefore, the measurement of visitor satisfaction with facilities and services is vital to the understanding of visitors, their needs, wants and willingness to pay (Fig. 7.4).

The concept of satisfaction has two key components. The first is a person's expectations. The second is their enjoyment of the experience. One's satisfaction is determined by how one's enjoyment exceeds the expectation, as shown by the following equation.

Satisfaction = Enjoyment – Expectation

Satisfaction equals enjoyment minus expectation. If one's enjoyment of a park, a service or a programme exceeds or equals expectation, then one is satisfied. However, if one's enjoyment is less than expectation, one is less than satisfied. Therefore, satisfaction is dependent on suitable levels of expectation and corresponding levels of enjoyment. Expectations are dependent on many factors. The service or facility levels experienced at one park provide an experience level that will be used in the future to create expectation levels. People often bring expectation levels from other situations; for example, the ability to register in advance for hotel rooms may lead to an expectation for advance registration for campsites. Not all people have the same needs. Some people want a car campsite for their tent trailer. Others desire a remote wilderness campsite for a tent. Some people expect a service or facility at a certain time, but not at others. Other

Fig. 7.4. The measurement of visitor satisfaction with facilities and services is vital to the understanding of visitors, their needs, wants and willingness to pay. Specialized activities, such as nature photography, often have unique needs. Park visitors often have high levels of knowledge and skill. It is important that managers collect information from all visitor types about their activities and satisfaction with facilities and programmes. Park visitor photographing white-faced capuchin, Manuel Antonio National Park, Costa Rica. (Photographed by Paul F.J. Eagles.)

factors include the appropriateness of the service in this type of park and importance of this service to different visitor types. Each factor in satisfaction must be looked at and evaluated in the context of the others.

Parks are often weak in their systematic measurement of client satisfaction. Parks that do not undertake frequent, obvious and valid satisfaction measurement force their visitors to become complainers. As a result, the only options available for such visitors to make their opinions known are to lodge a complaint, to disparage the park to others, to decide never to return or all of the above. Even in the absence of sophisticated client satisfaction measurement, keeping a good record of the complaints provides managers with useful information.

An example of the use of camper satisfaction measurement is from Ontario Parks, the agency responsible for Ontario provincial parks in Canada. This agency has a policy of surveying a sample of visitors in each park once every 6 years. The surveys are detailed and long. Fortunately, park visitors understand the need for such data and readily fill out the forms. In 1996, 11,834 campers completed the survey in 44 parks.

One section of the questionnaire asked campers to rate 25 facilities and services provided by the parks. This section was designed to collect importance/performance data; that is, how important a facility or service was to each person and how the park's performance was in delivering this service. For both sets of ratings, a scale of 1 to 5 was used. In the importance ranking, a rating of '1' indicated that the facility/service was considered 'Extremely Important', while a rating of '5' indicated the facility or service was 'Not at All Important'. In the performance ranking, a rating of '1' indicated that the park facility/service was considered to be 'Excellent'. A ranking of '5' suggested the performance of the park in delivering the facility or service was 'Very Poor'. Therefore, the closer the numerical rating on either scale to 1, the more important the facility or service or the better the park performed in delivering the facility or service.

The significance of this numerical form of data collection is that the importance can be compared with the performance mathematically. If the importance exceeds the performance, then the park is seen to have an importance/performance gap. Table 7.2 shows the importance/performance outcome of this large survey from Ontario Parks in 1996.

On a scale of 1–5, all importance measures under 3 are in the 'Average' (3) to 'Excellent' (1) rating. Clearly, Ontario campers feel all of the listed park services and facilities have a high level of importance to the campers. The most important was 'cleanliness of washrooms', while the least important was 'availability of heritage education programmes'.

However, the performance scores tell a different story. The highest performance was 'helpfulness of staff', while the lowest was 'availability of groceries/supplies'. Importantly, overall, the performance scores fell in the 'Above Average' to 'Excellent' range (i.e. 1.79–2.85), indicating that, in general, campers felt the parks' performance was very good for each of the 25 facilities and services rated.

Table 7.2. Ontario Provincial Park campers' revealed importance/performance gap (data from Stevor Inc., 1997).

Park service and facilities	Importance means[a]	Performance means[a]	Gap[b] (I – P)
Cleanliness of washrooms	1.47	2.21	–0.74
Availability of showers	1.77	2.26	–0.49
Level of privacy	1.70	2.14	–0.44
Condition of facilities	1.68	2.08	–0.40
Condition of campsite	1.61	1.95	–0.34
Control of noise	1.75	2.03	–0.28
Value for fee	1.80	2.05	–0.25
Availability of groceries/supplies	2.62	2.85	–0.23
Sense of security	1.60	1.81	–0.21
Condition of beach	1.78	1.99	–0.21
Control of pets	1.88	2.09	–0.21
Quality of firewood	2.25	2.46	–0.21
Recycling initiatives	1.90	2.02	–0.12
Responsiveness of staff to concerns	1.73	1.84	–0.11
Adequacy of signs	2.03	2.10	–0.07
Helpfulness of staff	1.77	1.79	–0.02
Condition of trails	2.04	2.05	–0.01
Ease of access to services	2.13	2.14	–0.01
Ease of campground registration	1.86	1.85	+0.01
Availability of information	2.07	2.04	+0.03
Recreational things to do	2.18	2.13	+0.05
Upkeep of park roads	2.17	2.09	+0.08
Availability of firewood	2.19	2.10	+0.09
Quality of picnic areas	2.35	2.08	+0.27
Availability of heritage educational programmes	2.68	2.33	+0.35

[a]Higher importance and performance are indicated by numbers closer to 1, on a scale from 1 to 5.
[b]The I/P gap is calculated by subtracting the performance mean from the importance mean. Therefore, a negative number means that the performance is below the importance assigned by the visitor. This indicates the need for management action on the park service or facility. A positive gap means the management performance exceeds the importance assigned by the visitor.

The importance/performance gap is an excellent measure of client satisfaction. The camper's expectation can be measured by the importance assigned to the factor, while the park's performance in fulfilling this expectation is measured by the stated level of performance.

Table 7.2 shows the park's services and facilities ranked according to the importance/performance gap, known as the I/P gap. Table 7.2 shows that only seven of the parks' services have a positive I/P gap, all the other performances were below expectations. Very importantly, the most important service of all, washroom cleanliness, had the highest negative

I/P gap (–0.74). Clearly, this agency must do a better job of washroom maintenance. This type of analysis is very useful. It provides clear evidence about where the Ontario Provincial Park managers should place their emphasis in the future. It also suggests where to place the emphasis when the park justifies higher fees.

Impacts of the Visit

Park visitation can have social, economic, cultural and ecological impacts. Specific studies by specialists are needed to properly assess such impacts. Larger parks and those with a higher public profile often prepare a series of studies that elucidate the impacts of park visitation. These studies direct management policy and provide important information to key stakeholders interested in park policy.

It is beyond the scope of this chapter to provide details on tourism impact studies. The important point is that such studies are essential if the full impact of tourism is to be understood and appropriately handled in park policy discussions.

Reporting of Results

The public data must be made available for the use of all those who require it. This includes government staff, such as park managers, tourism planners and local municipal authorities. It also includes the private tourism industry whose business life depends on a good knowledge of use levels and trends. Park visitors can use the data to help plan their visit. Interest groups, such as environmental organizations, use the data to better understand the level of use and impact of this use on park resources.

The data must be in forms that are clear and understandable. It is very useful to have it presented so that trends can be discerned. For example, presenting yearly data compared with previous years' data can be very useful. It must be presented in such a way that specialists can clearly understand how the data were collected and tabulated.

The public use data need to be provided to park managers on as frequent a basis as management requires. Some of it must be immediate, such as the current capacity utilization of the campground. Some can be medium term, such as the level of permit sales at the gate for the last month. Some can be longer term, such as total number of people utilizing the park interpretive programme over the last year.

Typically park agencies publish reports showing public use data over extended periods of time, usually on a yearly basis. These data are made available to a broad audience, often many months after collection.

Recently, public and private tourism industry demands requested park agencies to provide data in a more timely and easily accessible manner. Some agencies now provide the data relatively quickly on Internet web sites and through e-mail distribution.

It is critically important that all users of the data have a good understanding of the definitions used, the methods used for collection, the accuracy of the data and the approaches to calculation. There can be serious inaccuracies in park use data, owing to weak data collection and analysis approaches. Parks Canada managers reported high levels of confidence in only 12% of their public use data. For example, in one situation the public use levels reported varied from 1 to 2.6 million entrances due to malfunctioning traffic counters (Praxis, 2000). As a result, this agency launched a programme to upgrade the public use management programme.

The Role of Audits

If public use data are to be used with maximum confidence, it is important that all data collection, tabulation and presentation procedures are as accurate as possible. In parks with intermediate-to-advanced public use measurement systems, one park employee has the responsibility to check all procedures for accuracy. However, there is justification for an external audit of important figures. An audit is an independent evaluation, undertaken by a neutral party. For example, a company's shareholders hire an auditor each year to ensure that all the data coming from the employees are valid, accurate and complete. Only with the audited data can stockholders be sure of what they really own and the financial health of the company. The same concepts are useful in public agency data. All members of the public, the owners of parks and protected areas, can only be sure of the condition of the public estate when valid data are available. Therefore, independent audits of public use data are important.

There are reasons why public use data might be suspect. Agency employees may be too busy to collect valid data; however, for personal reasons may portray unrealistic images of data accuracy. It is not unknown for staff members to creatively fabricate data, to cover mistakes or to mask laziness. If park staff members feel that there is value in having higher or lower data than in reality, they may change the data to further their political agendas. It has been known for park managers to inflate their public use figures in order to try to impress their political masters of their success in attracting visitors and their need for a higher budget. For all of these reasons, it is necessary to have independent audits of park use data.

Competent park agencies employ internal auditors to check all park data, from budget figures through to public use information. In some countries, government auditors from outside the agency periodically audit park data. Very rarely, and typically only after some emergency has taken place, auditors are used from outside government.

Fig. 7.5. Monitoring revealed that beach users were damaging the sensitive sand dunes behind the beach. As a result, park managers built a large boardwalk over the dunes and separated the beach from the dunes with a fence. This action enabled existing levels of public use on the beach to be maintained while also protecting dune features. Beach and dunes, Kouchibouguac National Park, Canada. (Photographed by Paul F.J. Eagles.)

Summary

The planning and management of a park or any other type of protected area is dependent on reliable information that is appropriate for the decisions at hand. The design and operation of a public use monitoring programme is a critical component of park management. The park must maintain the findings of the monitoring programme so that future studies can benefit (Fig. 7.5).

This chapter provides basic concepts in such a programme, but only touches on many of the important details that make such a programme functional. We argue in this chapter that the carefully designed and well-operated public use measurement programmes require a level of dedication and sophistication that in turn requires strong management emphasis by park administrations. We also point out that all users of such data should be familiar with the accuracy of the data collection programmes before using such data for decision making.

References

CEQ (Council on Environmental Quality) (2001) *The National Environmental Policy Act of 1969*. Homepage of the Council on Environmental Quality, online. Available at: ceq.eh.doe.gov/nepa/regs/nepa/nepaeqia.htm

Hornback, K.E. and Eagles, P.F.J. (1999) *Best Practice Guidelines for Public*

Use Measurement and Reporting at Parks and Protected Areas. World Conservation Union, Gland, Switzerland. Also found at: www.ahs.uwaterloo.ca/rec/worldww.html

IAIA (International Association for Impact Assessment) (2001) *Welcome.* Homepage of the International Association for Impact Assessment, online. Available at: www.iaia.org/

NPS (National Park Service) (2001) *Frequently Asked Questions About The National Park Service.* Homepage of the National Park Service, online. Available at: www.nps.gov/pub_aff/e-mail/faqs.htm

Praxis Inc. (2000) Review of Priorities for Social Science Research within Parks Canada. Unpublished reported submitted to Parks Canada, Hull, Quebec.

Stevor Inc. (1997) The 1996 Ontario Parks Camper Survey: an Analysis of the Facilities and Services. Unpublished report submitted to Ontario Parks, Peterborough, Ontario.

Watson, A.E., Cole, D.N., Turner, D.L. and Reynolds, P.S. (2000) *Wilderness Recreation Use Estimation: a Handbook of Methods and Systems.* General Technical Report RMRS-GTR-56. US Department of Agriculture, Forest Service, Rocky Mountain Research Station, Ogden, Utah.

CHAPTER 8
Tourism Services and Infrastructure

Introduction

Parks and protected areas require facilities and services to accommodate use by visitors, park staff and others. Park tourism requires facilities for information, transport, accommodation, food, safety and recreation. To fulfil these common needs, facilities provided may include visitor centres, roads, campgrounds, grocery stores, water and sewage facilities and trails. The type, size and design of facilities established for an area are dependent on the objectives of the site. A wilderness park will have quite different tourism service and infrastructure needs from a national historic battlefield.

The design and operation of park tourism services is a specialized field. The special demands of unique ecological and cultural sites require care in the provision and management of facilities. This chapter provides a summary of the factors involved in the design and operation of tourism services and infrastructure. The goal is to provide the reader with an understanding of the importance of such design and the need for its careful application in parks.

Functions of Park Facilities

Facilities in parks and protected areas are designed to fulfil basic functions (Table 8.1). Facilities enhance a site's ability to provide recreation opportunities. A site with transport access provides higher levels of use opportunity than a site without such access (Fig. 8.1). A natural or cultural site often has limited ability to handle recreational visits. Basic facilities such as a water supply and washrooms are essential. Such facilities increase a site's ability to handle visitation. All sites provide some degree of danger to the visitor. This can vary from the trivial, such as an uneven walking surface, to the profound, such as an erupting volcano. Park facilities such as walking trails, barriers on exposed cliffs and rescue equipment provide various levels of public safety. Properly designed park

Fig. 8.1. Park tourism requires facilities for information, transport, accommodation, food, safety and recreation. Transport infrastructure is critical to park visitation. The placement and design of the transport facilities affect the environment and the visitor experiences profoundly. Highway 60 in Algonquin Provincial Park, Canada. (Photographed by Paul F.J. Eagles.)

Table 8.1. Functions of park facilities and infrastructure.

Functions
Enhance recreation opportunities
Increase capacity for tourism
Support visitor safety
Decrease maintenance costs
Be fiscally responsible
Protect and maintain ecological integrity
Support for cultural integrity
Maintain the health of air, water and soils
Provide for park staff needs

facilities can decrease maintenance costs. For example, a properly designed road will have much lower maintenance costs than one that requires constant repair. All facilities must be fiscally responsible. They must operate at minimum costs and should assist the park in its fiscal responsibilities. For example, a park entrance road and gate facility can guide visitors into a fee collection area, thereby increasing the park's efficiency in collecting use fees.

Parks have staff. In order to facilitate their work, many facilities are needed, ranging from maintenance buildings to office space. If the staff members are to be happy in their work, some of their private needs may need to be accommodated in the park. Many remote parks provide specialized housing for the staff who must live away from home for long periods. Others provide communities for the staff so that their family life can be accommodated in remote areas.

Most parks have ecological integrity goals. The maintenance of air, water and soil quality is necessary. Many parks have cultural and historical goals. The maintenance of cultural value of sites, artefacts and structures is necessary. The park facilities and infrastructure must be designed in order to support these goals.

Sensitive Design of Tourism Facilities and Programmes

Many older park facilities were designed to urban standards. Increasingly, more emphasis is being placed on sensitive design of tourism facilities and programmes. One of the earliest and most comprehensive works produced by the National Park Service of the USA (NPS, 1993) states that: 'Sustainable design is the philosophy that human development should exemplify the principles of conservation, and encourage the application of those principles in our daily lives.'

Key components of sustainable design of facilities include: maintaining biological diversity, supporting cultural integrity, buttressing the health of air, water and soils, reflecting local ecological conditions, producing minimum negative impacts, utilizing minimum levels of resources, having fiscal responsibility and supporting visitor safety (Table 8.2). Each

Table 8.2. Key components of sustainable design.

Maintenance of biological diversity
Support for cultural integrity
Maintain the health of air, water and soils
Reflect local ecological conditions
Produce minimum negative impacts
Utilize minimum levels of resources
Be fiscally responsible
Support visitor safety

of these components must be considered during the design and operation of all aspects of a park, with particular importance for park buildings and other forms of infrastructure.

The key components of sustainable design are applicable to nine aspects of park operations (Table 8.3). These are: interpretation programmes and facilities, natural resources management, cultural resources management, site design, building design, energy management (Fig. 8.2), water management, waste prevention and management, and facility maintenance. The National Park Service of the USA provides detailed recommendations on sustainable design in all nine areas (NPS, 1993).

Fig. 8.2. The use of renewable energy sources is desirable in many park situations. This solar energy station in Point Pelee National Park provides power to a remote area of the national park. The station also provides interpretation information for the instruction of the visitor about the potential of solar electricity generation. Solar energy generation station in Point Pelee National Park, Canada. (Photographed by Paul F.J. Eagles.)

Table 8.3. Park management areas requiring sustainable design.

Interpretation
Natural resource management
Cultural resource management
Site design
Building design
Energy management
Water management
Waste management
Facility operations and maintenance

Code of Sustainable Practice for Park Tourism

Tourism Council Australia adopted a Code of Sustainable Practice for Tourism (TCA, 1998). These principles, which were developed for all tourism, are also relevant to park tourism (Table 8.4). They are discussed in adapted format below.

1. *Use Resources Sustainably.* The conservation of resources is crucial in tourism. Natural, social and cultural resources are all important. Their conservation is important for society and for long-term business success.

2. *Produce Minimum Levels of Waste.* Minimum waste production reduces environmental damage, contributes to tourism quality and reduces financial costs in the long term.

3. *Maintain Diversity.* Maintaining and promoting natural, social and cultural diversity is ethically necessary and provides a resilient base for a long-term tourism industry.

4. *Integrate Tourism in Planning.* It is necessary for tourism to be a component of all national, regional, local and park planning. Such integration makes for a stronger tourism industry and results in much more coherent development policies.

5. *Support Local Economies.* Tourism that supports local economies is stronger politically. It is better accepted by local people and is more sustainable over the long term. Parks that support local economies have much higher levels of local public support.

6. *Involve Local Communities.* The full involvement of local people in decision making, in service provision and in tourism operation produces a population that understands and supports tourism.

7. *Involve Indigenous Communities.* Many parks are located in remote areas with aboriginal populations. Many such communities are interested in involvement in tourism. When involved, the people are more likely to understand and to support tourism.

8. *Consult Widely.* Widespread consultation with interested stakeholder groups and interested members of the public is essential if the tourism

Table 8.4. Code of sustainable practice for tourism.

Use resources sustainably
Produce minimum levels of waste
Maintain diversity
Integrate tourism in planning
Support local economies
Involve local communities
Involve indigenous communities
Consult widely
Train staff
Market tourism responsibly
Undertake research

activity is to be sustainable socially and economically. Such consultation can avoid conflict and assist with conflict resolution.

9. *Train Staff.* Suitable levels of staff training are essential for the success of all sustainable tourism practices. Such training also improves service quality, customer satisfaction and financial viability.

10. *Market Tourism Responsibly.* The development of a tourism market that is appropriate for the available resources is critical. Having the right tourists, with the right expectations, at the right place and at the right time assists immensely in the operation of sustainable tourism.

11. *Undertake Research.* Research and monitoring is essential to understand environmental, social and cultural impacts, long-term trends and market changes.

Unfortunately, park tourism programmes are seldom evaluated to assess whether they follow sustainable principles such as those outlined above. Therefore, it is not possible to report the degree of implementation based on systematic data. It is clear that increasing numbers of public and private tourism operations in parks are becoming aware of such principles and attempting their implementation.

Impact Assessment

All tourism has social, cultural and environmental impacts. Park planners purposely design for some of these impacts. For example, a park tourism industry may be developed to provide a local community with positive economic impact. Or a park may be created to encourage visitors to observe and gain appreciation of a unique ecosystem. Some impacts may not be so desirable, such as local people's emulation of tourists' lifestyles or campground sewage changing the chemistry of local water supplies. Therefore, all aspects of park tourism must be carefully assessed for their social, cultural and environmental impacts. Such analysis is typically done before programme and facility design. Once a facility is constructed and a programme operated, monitoring data provides insight into its success. Therefore, impact assessment is done during both the design and the operational periods.

The determination of the value of the impact is critical. Any impact can be considered to be negative, neutral or positive depending on several factors, such as the point of view of the observer, the time of year or the costs and benefits derived. Impact assessment is fundamentally a political, not a scientific, process.

The determination of value is a major part of impact assessment in parks. There are many key groups that play roles in the determination of impact value. Who assigns value and how the value is assigned are central issues. Also critically important is the method used to assess this value.

In practice, many constituencies influence park management. Natural and cultural resource decision making takes place within this context of many values and many interested groups. The decision-making system must realistically and effectively provide an opportunity for most people in most constituencies to participate. No one group should be allowed to dominate.

The soundness of all decisions is heavily influenced by the values and knowledge of the involved public. It is important that the ecological and cultural roles of parks are communicated to and understood by the full range of publics that influence decisions.

In reality, every major decision in parks is ultimately subject to a formal political process. How many people support a particular decision is the telling point for democratic governments. All park managers and all public stakeholders must be aware of this fact.

The assessment of impact is dependent on the determination of value within an overtly political process. The park visitors are one of the least influential groups in decision making. They visit for a short period, often live far away from the park and are poorly organized politically. It behoves park officials to develop procedures to ensure that park visitors and potential park visitors are given a voice in park decision making. Since the park visitors are the major beneficiaries of park facilities and programmes, it is necessary for decision makers to provide opportunities for these people to influence the design and the operational policies of the park.

Design and Maintenance of Park Infrastructure

All parks require some level of infrastructure. This can usually be categorized into three groups according to the types of users: to serve the needs of park employees, of visitors and of others. The first category fulfils the basic needs of the park employees. These people require places to live and work. The fulfilment of their responsibilities often requires specialized facilities such as roads, trails, air strips, laboratories, equipment maintenance buildings, storage depots, offices, communication towers and remote cabins. The second category fulfils the needs of the park visitors. Typically, these include accommodation, facilities for purchase of supplies, restaurants, equipment rental, facilities for search and rescue, safety, sanitation, water and sewage supply, and transport. Fortunately, the visitors and the park staff can both use many of the same facilities. The third category is an eclectic group that includes all people that are neither staff nor park visitors. These can include local people with specific rights of access, such as for the collection of natural resources. Large parks may have transport facilities for people who just pass through. For example, Banff National Park in Canada has the Trans-Canada Highway running through the park as well as major railway tracks. This park must accept and manage very large levels of infrastructure and impact from transport facilities that satisfy needs generated well away from the park.

It is also possible to understand park infrastructure through the classification of types of use: recreation, tourism support and infrastructure. There is a large body of literature and experience dealing with the design, construction and maintenance of outdoor recreation facilities (Fig. 8.3). They range from the relatively simple design of a nature trail, through the more complex design of campgrounds to the highly technical engineering involved in cable-car systems and mass transit facilities. The basic tourism support facilities include roads, stores and restaurants. Basic infrastructure includes that for communication (telephone, radio, television, Internet), safety (barriers, signs, walls), water supply (wells, reservoirs, treatment, storage, distribution), sewage (collection systems, treatment, disposal) and waste (collection, storage, movement, landfills, shipping).

There are three generally recognized levels of expertise needed for park infrastructure. The lowest level is that within the capability of park labouring and maintenance staff. These can include simple projects such as placement and maintenance of signage, campsite maintenance and waste collection. All parks have a requirement for management and maintenance staff to maintain the park basic facilities. Labouring staff with the ability to fix and maintain a wide variety of infrastructure are highly desirable. The more advanced level involves people with specialized training in design, such as planners and landscape architects. These people design trails, campgrounds and basic facilities. Very few parks have such staff, but many park agencies maintain planning and design staff to

Fig. 8.3. Over time, park managers develop expertise in the design, construction and management of specialized outdoor recreation facilities. The heavy rains and winds of the New Zealand southern mountains require special facilities for the visitors. This shelter in Mt Cook National Park provides shelter for visitors. Shelter in Mt Cook National Park, New Zealand. (Photographed by Paul F.J. Eagles.)

develop management and site plans in many parks. The most advanced level involves specialists in fields such as engineering and architecture. This advanced level is required for facilities such as roads, major buildings, water distribution and sewage systems. Only the largest park systems have permanent staff in engineering and architecture. Generally such expertise is purchased from the private sector when required.

All infrastructure can have major environmental, scenic and cultural impact. Therefore, it is important that parks design, construct and maintain such infrastructure in a manner that is environmentally, culturally and financially sensitive.

Energy, Water and Waste Management

Facilities and service provision in a park generally require energy, water and waste management. The provision of energy, water and waste management not only strongly affects the quality of the visitor's experience, it often consumes large parts of park budgets. In this section we provide an overview of the principles involved in the selection and use of various approaches to such management.

Energy management

The availability and use of energy is a vital element in tourism. The import and use of electricity, petrol, diesel fuel and camp fuels, such as kerosene, are widespread in parks. There is a strong desire for parks to show environmental leadership by the use of as little energy from non-renewable energy sources as possible. There are three main types of renewable energy sources: wind, solar power and hydropower. Table 8.5 compares these three types. These sources can only be used when there is sufficient wind, sunlight and water. Solar power has considerable advantages over wind and hydropower. These advantages are becoming larger as higher efficiency electrical generation cells are developed. However, both wind and solar power must be stored during periods of low generation. Hydropower has the biggest environmental impact owing to the construction of dams and water storage impoundments. However, once the water is stored, it is readily available for electricity generation at any time. In all three cases, the issue of scale is important. Small-scale generation facilities typically have a lower local negative impact and footprint on the landscape than do larger-scale facilities.

Parks are increasingly utilizing renewable energy sources (Fig. 8.4). The remoteness of many parks and of many park facilities makes the utilization of such energy sources attractive. When utilized, many parks provide interpretive material to the park visitors emphasizing this

Fig. 8.4. Many parks show environmental leadership by the use of energy from renewable and non-polluting energy sources. Park interpreters in Kruger National Park use an electric vehicle to provide quiet, unpolluting transport for park visitors during evening game drives. Electric vehicle for evening game drive in Kruger National Park, South Africa. (Photographed by Paul F.J. Eagles.)

Table 8.5. Renewable energy sources compared (adapted from Boele, 1996).

	Wind energy	Solar power	Hydropower
Cost of installation	Medium	Low	High
Site-specific conditions	Medium	Low	High
Seasonal variations	Medium	Very high	Medium
Running costs	Medium	Low	Low
Noise	Yes	No	Low
Reliability	Medium	High	High
Capital cost	Medium	High	Varies

utilization and encouraging the application of such technologies to the visitors' personal lives.

In many parks, readily available local energy, a lack of technical expertise in renewable energy management and the capital costs of alternative energy generation facilities result in heavy utilization of traditional energy generation sources, such as from local electrical grids and diesel generators. It is probable that increasing operating costs of non-renewable energy in the future will push many parks to utilize renewable energy to a much higher degree. This trend will be further emphasized as the costs and efficiencies of alternative energy generation, such as solar generation, become more attractive.

In some special situations, parks require the use of non-renewable energy by visitors. In high altitude situations, many parks prohibit the use of renewable energy supplies, such as wood, due to the negative impact of local firewood collection. In these situations all cooking and heating must be done with liquid fuels. The decision here is that in the conflict over two undesirable practices, utilization of fossil fuels and burning of scarce local wood, the one that has lower levels of local negative environmental impact will be used. However, it must be recognized that the impact caused by the collection, refining and transport of such liquid fossil fuels still occurs, but at a site remote from the park.

Water management

Water is needed by people in parks for consumption, cleaning and waste management. Urban people often expect sizeable amounts of clean, fresh water available for their use. Sometimes this expectation cannot be fulfilled. In parks with scarce water supplies, the provision of water to people involves several important principles. First, the users are informed of the scarcity of water and the need for minimum use. Secondly, all water facilities are specially designed for minimum water use. Items such as composting toilets and automatic shut-off taps are commonplace. Some sites use advanced technologies to recycle water. Thirdly, interpretive programmes provide visitor education on the technologies involved in sustainable water utilization.

Waste management

Waste management is a major issue in parks. People typically generate high levels of waste, often with negative impacts. Common categories of waste include sewage, litter, plastic, glass, paper, metals and organics.

Human sewage requires special consideration. It can pollute water bodies and affect ecological conditions. The management response can vary from high-technology sewage plants for high-volume sites through to individual organic toilets. In remote situations, low volumes can be handled with individual waste burying or by primitive toilets. Many parks took decades to develop suitable sewage management procedures. Even today, sewage pollution is a problem in some parks.

Litter is the distribution by visitors of small amounts of waste along trails, at recreation sites and near car parks. Litter collection programmes, visitor education and enforcement of anti-litter regulations are commonly used to reduce litter. Some parks are utilizing special programmes of litter collection by visitors as a method of involving interested people in park management. The visitors are rewarded with recognition and special items of acknowledgment, such as badges and citations.

Normally, plastic, glass, paper and metal wastes are collected and placed into landfill sites. More recently, tourists familiar with recycling programmes at their homes have demanded that parks introduce waste collection and recycling programmes for plastic, glass, paper and metals. These programmes can be very expensive for park authorities because of the cost of transport of these products to remote urban collection points. Nevertheless, sustainable ecological principles demand that parks function in the most ecologically sustainable fashion possible, and are seen to be doing so.

Food waste can create novel impacts. When collected from tourists and placed in landfill sites, interesting impacts often result. Landfills are attractive to wildlife and sometimes provide wildlife viewing opportunities. Most parks now block access to the landfills in order to keep park visitors from utilizing these sites for wildlife observation. Some species gain sufficient resources from the waste sites to increase survival rates and thereby inflate populations. These populations then have a cascading effect on the population of related predator and prey species. To alleviate such problems, most parks have moved their landfill sites outside the parks and have developed programmes to reduce waste generation.

In a campground, food waste attracts wildlife. Species that utilize such waste may increase in numbers to undesirable proportions. Dangerous wildlife may enter the campgrounds to gain access to human food. Some parks have developed novel solutions to this issue.

The costs of improper food handling, waste collection and disposal can be considerable. For example, the wilderness canoe routes in the interior of Algonquin Provincial Park in Canada have about 250,000 visitor nights of use per year. All the remote campsites can only be reached by canoe. The waste generation from such a large number of visitors in such a remote location became a major management problem. Visitors were required to bring out waste. However, a small amount of non-compliance led to an expensive waste problem. The collection of waste from thousands of remote sites scattered over thousands of square kilometres was time consuming and financially prohibitive. Park management looked for a solution to the problem and introduced a can and bottle ban for all interior users. All supplies, including food, were required to come in containers that were burnable, biodegradable or reusable. This park regulation in turn stimulated innovation in the production of dried foods and reusable containers for all the products used on canoe trips, ranging from stove fuel to food. A large industry developed to serve this market for lightweight products in sustainable packaging. The management decisions, the technological innovation and the universal acceptance of the policies by the wilderness canoeists led to an acceptable solution to the problem. Associated with this effort was an education campaign to prepare wilderness campers for dealing with bears who might enter campsites to access food. This campaign involved many elements, such as proper food storage on the campsite, proper waste handling and education about bear behaviour. This campaign was successful in reducing, but not eliminating, negative interactions between interior campers and black bears.

Waste management is an expensive aspect of park tourism management. Many park visitors are not aware of the financial costs of such management and are not therefore amenable to paying appropriate levels of fees for such management. In such situations, the park must aggressively inform its visitors of the waste management situation, of the costs involved and the needs for sufficient finance to cover these costs.

Best Practice Principles of Sensitive Tourism Facility Design

The high levels of interest in sensitive tourism design lead to the search for good examples of its application. The Western Australian Tourism Commission undertook such a search, and published its findings (Crawford, 2000). To assess when best practice occurred, the nature-based tourism development can be evaluated with five criteria.

1. *Sustainable design* must be evident. Tourism must be designed with and for the environment. An ultimate goal is a unique sense of place of the programme and facilities.
2. *Interpretation* must lead to consumer awareness, appreciation and understanding of environmental processes.
3. *Local community involvement* must be strongly evident. This can occur with the direct involvement of local people and through the inclusion of local cultural elements. This can occur with the site's architecture and delivery of hospitality services.
4. *Financial return to the environment* must be evident. The tourism businesses must return a portion of their income to environment conservation.
5. *High quality* in food, beverage and other hospitality services must be present.

The Tourism Commission evaluated select nature tourism sites in Kenya, Tanzania, Zimbabwe, Belize and the USA. Other similar programmes include the yearly award programmes for sensitive nature tourism operated by Conservation International and British Airways. The formal evaluation of sensitive tourism design and the awarding of honours for success provide encouragement for the spread and application of such principles.

Summary

People in parks require facilities and programmes. The form of these facilities and programmes follows their function. Their functions depend on the objectives of the park and protected area. Therefore, the overall goals for the park and protected area are the umbrella under which all programmes and facilities are sheltered (see Fig. 8.5).

Fig. 8.5. The design and operation of park tourism services is a unique field. Specialized facilities provide exceptional visitor experiences. This highly engineered, but sensitively designed, treetop walkway allows visitors to experience the forest canopy of the giant red tingle trees of southwestern Australia. Treetop walk in Valley of the Giants, Walpole-Nornalup National Park, Western Australia. (Photographed by Paul F.J. Eagles.)

All facilities should be guided by the overall values underlying the park. These values include the historic values underlying the park's creation and the values of the park staff and the park visitors currently involved.

Park infrastructure is a critical component of the visitor experience. It must be carefully designed and managed in order to fulfill both the needs of people using the park and the protection of the environmental and cultural values of the park.

The application of sustainable and sensitive design principles to park tourism facilities and programmes is an important and growing field. Their use is stimulated by the idea that a park must utilize the best practice possible and must serve as a good example. Tourist pressure can be a powerful force to stimulate park agencies to adopt better practices. Park agencies and tourism bodies are gaining considerable experience in the provision of sustainable design and management of park facilities and programmes. Such experience is becoming available in published documents and on the Internet.

It can be expected that higher levels of sustainable design and operation will be used in parks in the future. Paradoxically, higher costs for energy and other natural resources can assist with the trend, but only if park management has sufficient capability to undertake the redesign and implementation.

References

Boele, N. (1996) *Tourism Switched On: Sustainable Energy Technologies for the Australian Tourism Industry*. Tourism Council Australia, Barton, Australian Capital Territory.

Crawford, S. (2000) *Designing Tourism Naturally: a Review of World Best Practice in Wilderness Lodges and Tented Safari Camps*. Western Australia Tourism Commission, Perth, Western Australia.

NPS (National Park Service) (1993) *Guiding Principles for Sustainable Design*. National Park Service, Denver Service Center, Denver, Colorado. Available at: www.nps.gov/dsc/dsgncnstr/gpsd/toc.html

TCA (Tourism Council Australia) and CRC Tourism (1998) *Being Green Keeps You Out of the Red*. Tourism Council Australia, Woolloomooloo, New South Wales.

CHAPTER 9
Tourism, Protected Areas and Local Communities

Introduction

One of the fundamental reasons to consider tourism within the context of national parks and protected areas is the linkage with communities that are either adjacent to or within the park. Tourism development has, as principal objectives, the creation and maintenance of economic opportunity, enhancement of quality of life and protection of a culture's historic and natural heritage. Indeed, much of the ecotourism and sustainable tourism literature speaks directly to responsibilities and opportunities to not only protect national parks through appropriate tourism development but also address many of the economic and social challenges facing local, and mostly small, rural communities. In many cases, these communities struggle with a changing economic base, one moving away from resource commodity production. Other communities face widespread low incomes that they desire to raise. For some communities that have traditionally relied on natural resources now gazetted within a national park or protected area, tourism represents an economic incentive to the community for protection of these areas.

For example, tourists visiting some African national parks often also visit local villages. The visitors may spend money on crafts, lodging, food and village entrance fees, thus enhancing the economic condition of the village. This result comes about only because of the presence of the nearby park. Without the park and its good management, visitors would not travel to these villages and spend money. Additionally, entrance and user fees to many national parks fund their needed management and maintenance activities. In Bwindi Impenetrable Forest and Mgahinga Gorilla National Parks of Uganda, a portion of visitor entrance fees goes to local villages and to finance park ranger salaries (Litchfield, 2001). For the community of Bermudian Landing in Belize, visitors to the nearby Black Howler Monkey Preserve represent added income for meals, lodging and guides. The entrance fees visitors pay also provide income to local farmers as compensation for protection of habitat that otherwise could be converted to agriculture. Such funds not only enhance the interests of local communities in the protected area (and its management), but they also increase the level of protection that can be given to park values (such as the endangered highland gorilla and black howler populations), thereby counteracting potential threats (such as poaching or habitat destruction).

The economic viability of a community may be the fundamental driver of its interest in protected area management (see Fig. 9.1), and often comes about because of structural changes occurring in the community's economic base. Such changes, frequently induced by governmental and corporate decisions made at a distance from the community, leave many towns and villages economically stranded. As factories and commodity processing plants close or reduce operations, communities search for alternatives to maintain the employment, income and tax revenue needed to provide governmental services. Tourism development is frequently perceived as a way that money can be brought into communities quickly without a relatively large capital investment in facilities and factories. Other communities may be fortuitously located adjacent to or along

Fig. 9.1. The economic viability of a community may be the fundamental driver of its interest in protected area management. Park tourism provides business opportunities for local people. This strawberry farm sells to the park visitors travelling to Volcan Poas National Park in Costa Rica. Strawberry farm near Volcan Poas National Park, Costa Rica. (Photographed by Paul F.J. Eagles.)

routes leading to a protected area; businesses may establish themselves along the route to provide goods and services needed by tourists. As the community understands its new linkage with a protected area, it takes interest in management policy.

Other factors also may tie a community to a protected area. Such linkages may be based on cultural or spiritual traditions; a protected area may contain sites or features that have become symbolically important to a community or its residents, such as a mountain peak, waterfall, gorge, lake or forest. In these cases, the community may have an equally strong interest in the protected area, but the interest may be in preserving the site and, in some cases, restricting non-local access to it. Management policies affecting such symbols may provide the basis for community–park administration dialogue.

The linkages between protected areas and communities thus occur at different scales – spatially, functionally and temporally – and, as tourism to protected areas increases, questions arise about these linkages, the role of tourism in them and how they should be addressed. In this chapter, we not only explore some of issues associated with tourism, communities and protected areas, but we also suggest some basic principles and processes useful in managing these linkages. We will also describe the character of negative impacts that occur to communities engaging in nature- or culture-based tourism.

Types of Tourism, Community and Protected Area Linkages

The links between communities and protected areas are often difficult to identify and understand. They occur at different temporal and social organizational scales (individuals, households, communities) and within differing functional areas (economics, spiritual, quality of life, etc.). In this section, we describe these linkages and the various considerations involved in them through several propositions.

The economic relationship between protected areas and communities is multidimensional

The economic foundation for community–tourism–protected area relationships involves at least four different linkages. In the first and most obvious linkage, visitor expenditures influence the economic viability of the local community. A community's economic base is often composed of a variety of industries and sectors: agriculture, manufacturing, services (financial, legal, etc.), mining, timber harvesting and processing, and tourism. The relative contribution of each to the economic base varies by community. A specific community may or may not be economically dependent on the tourism that occurs in a nearby national park or protected area.

The effect of visitor expenditures (direct, indirect and induced) is often what we refer to when discussing the economic impact of protected area tourism on communities. As we noted elsewhere in this book, such impacts can be a significant part of a local economy. In this sense, then, communities are often interested in increasing levels of visitation to a park to raise the associated visitor expenditures.

The community of Lukla in the lower Khumbu region of Nepal serves as the gateway for trekkers and climbers exploring the Mt Everest region of Sagarmatha National Park. The community did not exist until an airstrip was constructed to provide better access to the area. Its economic base is tied totally to the visitation that results, and, thus, increases in tourism result in more favourable economic conditions for the community's residents. Policies that affect air access to the village have direct consequences for its economic viability.

In addition, park and protected area administrations that are reliant on entrance and user fees to fund park management often seek increases in visitation as well, particularly when current administrative expenditures are funded from current fee revenue. For example, the Saba Marine Park, located in the Netherlands Antilles, funds its management and maintenance programmes totally from diving and other fees associated with use of the park. When changes in visitation occur, there are direct consequences to the ability of park management to fund necessary activities. Declines in visitation may lead to strategies to counteract the trend to increase revenue streams. Unfortunately, this strategy may be in direct opposition to the needs to manage and control the spread of negative visitor impacts.

The second linkage involves the protected area and its landscape, which serve as a scenic backdrop to adjacent communities. Such scenic backdrops serve as a component of a local area's quality of life. The community of Bozeman, Montana, in the USA is located adjacent to several national and regionally significant protected areas, including Yellowstone National Park. Since the early 1980s, firms and individuals seeking to work and live in a place that has easy access to protected areas and scenically attractive landscapes surrounding the community have enhanced its economic viability. Thus, the park may be an important resource not only for residents to enjoy but also in terms of economic development activities; because it is such an attractive component of the community, firms and businesses desire to locate there. While such businesses may not have a direct relationship with the protected area or its resources, the employees may use the park as a place for recreation, without which they may not have moved into the area.

Thirdly, the expenditures of the park or protected area itself serve as part of the economic base for the community when funding for park management and development activities comes from outside the community through fee revenue or central governmental appropriations. The park administration may then purchase goods and services needed for management within the local community. Such expenditures are a net increase in revenues to the community because the funding for park administration comes from outside the community. Depending on the size of the park budget, these expenditures may be substantial in themselves.

Fourthly, the employees of the park and private tourism businesses reside in the community and spend their salaries and wages on needed goods and services within the community. Employee expenditures may be sizeable, and often communities and the businesses within rely heavily on such spending. Such secondary spending may also stimulate additional spending on goods and services within the community by receiving firms.

As a result of these four types of impacts, communities often have a very strong interest in how a park is managed and the consequences of management actions not only to the local tourism industry but also to the entire community itself. Changes in management, development of new facilities and shifts in programme emphasis thus affect a community's economic relationship with a park or protected area. In some cases, such changes create uncertainty. The resulting anxiety over how a change in park policy affects a community and the individual businesses within it often becomes a source of conflict between the community and park administration. In addition, the four types of economic impacts indicate that communities have strong business and financial interest in a park, and will be expected to respond to any proposed changes from a strictly economic perspective.

The protected area may hold values important for the residents of the local community

In addition to the economic value that a park or protected area may hold for an adjacent community, there may be other socially important values contained within the protected area. For example, a protected area may contain resources important for traditional subsistence uses such as forage, plants or animals. In the UK, national parks are designed to protect mainly human-modified rural landscapes. In this case, objectives of park management are quite explicit about maintaining traditional pastoral practices and landscapes. These practices are emphasized over those that would restore landscapes to a natural condition. In other cases, such as in many of the National Reserves located in the state of Alaska in the USA, indigenous peoples are permitted to maintain subsistence fishing and hunting activities. In both examples, protected area managers must understand that the linkage between the community and the park is utilitarian, if not necessarily economic, as defined above. Continuation of traditional agricultural and subsistence practices is therefore of interest to local communities.

Communities often have deeper and more spiritually based linkages to resources and features contained within a protected area. Important historical events may have occurred within the park boundary. Such events may be culturally important to an adjacent community and thus the community will have a strong interest in protecting a particular site and ensuring that its surrounding environment maintains a high level of historical integrity. For example, management of the protected Civil War battlefields in the USA, such as at Gettysburg, emphasizes maintenance of the vegetative landscape and land uses as they were when the battle was originally fought.

For some individuals and communities, places within a park may retain or hold a spiritual significance. For example, Devils Tower National Monument in the state of Wyoming in the USA traditionally served as an important visioning site for some of the North American Indian tribes of the Great Plains. This spiritual characteristic is in conflict with newer uses of the monument that allow climbing and hiking to the top for recreational purposes. The Head-Smashed-In Buffalo Jump in the province of Alberta in Canada also preserves the place where native people of the mountain foothills drove American buffalo off the cliffs for more than 5500 years. Such spiritual and cultural characteristics would be unintentionally marginalized in any planning processes that only dealt with utilitarian relationships between communities and parks.

Communities often have a political relationship with an adjacent park

National parks and protected areas represent a national commitment to protection and preservation (see Fig. 9.2). This commitment may be

Fig. 9.2. Communities often have an important political relationship with a park. In this photo the most senior park official in Tanzania talks casually to local people, thereby keeping channels of communication open. This interaction takes place in a show village, set up to allow local people to display their culture and crafts to safari tourists passing between Ngorongoro Crater and Serengeti National Park. The Director of Tanzanian National Parks, Lota Melamari, meets with local men in Ngorongoro Conservation Area. (Photographed by Paul F.J. Eagles.)

viewed at the local level as being at odds with community interests. Such designation represents a political decision to preserve and protect certain landscapes, cultural resources and biodiversity values. But this decision, generally made at a national level, may be made without the full support or endorsement of a local community because designation may limit access to resources and areas traditionally used by community members. An example of such a situation is the Sagarmatha National Park in Nepal. When the park was created, there was a lot of antagonism towards it by people residing in local communities. Even though the local communities were not included in the park, the park was seen as an impediment to traditional ways of life. For example, Sherpa (1993) describes how members of local Sherpa communities feared that establishment of Sagarmatha National Park would lead to a reduction in traditional uses and rights to forests and grazing lands. Restoration of traditional communal management systems in some communities and more local–park interaction later served to reduce local anxiety and opposition to park management.

Individuals within the community may hold strong reservations about the designation itself; such reservations about designation may carry over to concerns about management. The superintendent, ranger or chief warden of such a park has a difficult job ahead in such situations. Much of this job would revolve around working with the local community to deal with its concerns, address its fears and reduce its anxieties. Unfortunately, fears and anxieties about designation and management often lead to feelings of distrust, creating a political environment in which communication is problematic.

Of course, not all communities necessarily have negative responses to proposals to designate nearby areas as national parks or protected areas. In many cases, such proposals are often viewed very positively by stakeholders within the community and by the community as a whole. Nevertheless, when two differing groups have interests in a particular idea or area, there are bound to be differences, thus establishing the need for both to cultivate their relationship.

Such political differences between local communities and national interests were the hallmark of protected area conflict in the 20th century. Often, managers sought to achieve a balance between the two interests. The notion of balancing interests implicitly assumes they are competing, that communities have an interest different from the national government. Often, however, there is much overlap in interests between communities and their national governments in protected area management. For example, communities, because of their economic linkage to a protected area, have an interest in its protection and proper management, as does the national government as part of its commitment to maintaining, at least symbolically, important landscapes and cultural heritage values. Thus, the question is not one of seeking balance but one of integrating or accommodating both interests.

In some cases, proposals to increase protection or reduce access may lead to short-term negative economic impacts on an adjacent community. These negative impacts may be a significant barrier to identifying mutual interests. For example, Yellowstone National Park in the USA is considering eliminating snowmobile use, thus adversely impacting the economic base of a local community, West Yellowstone, Montana, which has been heavily dependent on snowmobiling in the past. Such snowmobiling activity bolstered the community's sagging winter economy. Both the community and the national government share an interest in the park, but one interest is defined more in economic terms while the other is defined in terms of biodiversity values. Such conflicts typify many community–park relationships, but this does not necessarily mean that they are incompatible or competing. There should be ways of ensuring community understanding of important park biodiversity values that lie at the heart of interest in winter access to the park. At the same time, the National Park Service must recognize the significance of its activities and policies for economic conditions in the local community. By integrating both interests, creative solutions to apparently zero-sum conflicts may be generated. Searching for shared values, interests and meanings, followed by developing venues that encourage dialogue, learning and mutual fact finding are fundamental to integrating these interests.

Social Impacts Resulting from Park-based Tourism

As Machlis and Field (2000) note, a visit to a park involves more than a visit to a park. The visit entails travel to and from the park as well as visits and stays at communities and villages along travel routes and perhaps the communities adjacent to or within the park, even if the visits are relatively ephemeral ones used to purchase minor goods or services. In a sense, development of park-based tourism is designed to take advantage of such visits, and make them longer, with greater levels of expenditures. Such visits lead to a number of social impacts, some intended, some unforeseen. These impacts may be both positive and negative, depending on who is impacted, how and what standards those affected use to evaluate the impacts. Thus, a local tourism business owner may view visits positively in the sense that revenue to the business is generated, leading to greater profitability of the firm. But another local resident may view tourism-related traffic as contributing to congested roads, and thus may perceive tourist activity negatively. In this section we provide an overview of the types of positive and negative impacts on the social system that may occur with tourism.

Positive impacts

Much of this book is about the positive impacts of park-based tourism. These impacts can be briefly summarized as economic, as indicated elsewhere in this text, social, political and cultural. The primary positive social impacts deal with the enhanced ability of a community to take care of itself, to provide a place for its youngsters to seek productive employment, to increase the educational levels of its citizens and to provide affordable access to housing and health care. In a sense, these effects increase the community's capacity to adapt to changes imposed on it. This increased resilience of the community increases its opportunity to remain viable in the face of external policies and economic conditions.

Positive political effects include increased awareness of issues and participation in community decision making. This allows community members to make more enlightened decisions about its future. Positive cultural effects involve increased pride in local customs, traditions and rituals. The probability of these being preserved may increase as community members see increased interest in them from their non-resident visitors. In addition, there may be enhanced interest in protecting and sensitively managing the local natural heritage as the community sees that tourists often are interested in it. However, these positive impacts can come about only with careful planning and management of tourism activity. At a larger scale, tourism development may lead to greater cross-cultural communication and understanding, leading to a reduction in international tensions.

Negative impacts

The tourists that visit national parks, as we have seen, often bring with them negative impacts and consequences that are difficult to address. For the communities adjacent to national parks, such impacts tend to be social, cultural and economic in character. They raise questions about how such impacts are distributed, actions needed to mitigate them and who should pay for such impacts, as in national parks. They confront the recipient with difficult trade-offs, such as the acceptability of social impacts vs. the desirability of economic growth and stability.

Negative social impacts occur when residents are impeded from acting out their daily routine or achieving ongoing goals, such as when traffic congestion occurs. Social impacts also occur when tourists and tourism developments lead to increases in the crime rate and illicit drug use, and increases in the availability of some services, such as emergency health care. Cultural impacts happen when fundamental normative beliefs are not only challenged by the presence of tourists, but are changed, as when the dress and appearance of tourists are adopted by residents. These types of impacts are particularly dramatic in the developing world and in communities where agrarian and spiritual traditions dominate or where indigenous peoples reside. In this section, we provide an overview of the types of negative impacts that may be experienced by a community as tourism to nearby national parks or protected areas increases. The extent and intensity of each type of impact is influenced by a number of variables, including the amount, type and location of tourism facilities; the amount and behaviour of touristic activity; the rate of growth in tourism; the character and distribution of the community's social capital; and the institutional structure and environment for coping with the direct and indirect consequences of tourism. Each community must decide whether such potential consequences represent viable choices in return for the possible economic and social benefits from tourism. Some of these impacts are discussed below.

Commodification Commodification occurs when culturally important symbols, events and icons are bought and sold by entrepreneurs without the explicit permission of the culture that 'owns' them. In particular, commodification is an issue when non-resident-owned businesses appropriate cultural symbols only for the purpose of making a profit. While commodification is an issue in many places, it is heightened in areas occupied by indigenous cultures. In her book, Deborah McLaren (1997) provides an example of commodification involving an Indonesian burial rite. In this case, the burial rite was advertised as a traditional cultural event for tourists to view (and pay for). Commodification occurs when a cultural or religious ritual now becomes a consumer product bought and sold in the marketplace.

Exploitation Exploitation involves an unequal power relationship when two or more groups interact. Exploitation in this sense occurs when one of the groups is taken advantage of, economically or politically. For example, local individuals or communities are not adequately compensated for goods and services they provide to tourists. Other examples of exploitation include charging extraordinarily high prices for scarce goods and services to non-resident visitors and excluding local residents from community-level tourism development decisions.

Transformation Transformation is a process that occurs when local communities and indigenous peoples adopt the cultural, economic and political practices of non-resident visitors. This occurs quite frequently in the private sector, when national or international companies locate standardized facilities, stores and restaurants in culturally unique communities. This process leads to a new host culture, which may have little relationship to the original host community. While cultures are always evolving, transformation represents an infusion of foreign influences into that evolutionary process. The result is a gradual loss of the idiosyncratic character of a local culture and its dominance by foreign corporate symbols, icons and behaviours.

Competition As tourism increases in a local community, tourists begin to seek the same cultural, biophysical and recreational opportunities that many of the residents may have sought when they originally located in the area. This leads to competition for scarce resources and a feeling among local residents that they are being crowded out of their favourite places, whether they are bars, parks or other socially important locations. In some cases, these feelings are particularly significant. On the Big Hole River in the state of Montana in the USA, increased competition for fishing has led to prohibitions on non-resident fishing on the river during certain days of the week in order to permit opportunities for local residents to fish.

Essentially, the attributes that make an area an attractive place to live are often the same ones that make the community and adjacent area attractive to tourists. In some cases, competition leads to displacement of residents both from their favourite recreation places and from their homes. Competition may lead to increased housing costs as well as elevated prices for goods and services, such as groceries. Prices may become so elevated that local residents and the workers that provide the needed services for tourists can no longer afford to live in the community. This has happened in Jackson Hole, Wyoming, as tourism activity and development resulting from visitation to Grand Teton National Park and the surrounding area accelerated rapidly. With a limited supply of housing, residents and service workers were forced to move to nearby communities, particularly in the nearby state of Idaho, in order to afford housing.

Demonstration When residents of a local area begin to adopt the clothing and material goods of non-residents a demonstration effect occurs. This effect is most noticeable in indigenous communities where traditional clothing is given up for the kinds of clothing and articles that tourists bring with them. An example would be Amazon Indians wearing T-shirts with a 'Hard Rock Café' logo on them or Sherpa women in Nepal giving up traditional dress for jeans and shirts. Such demonstration effects partly occur because of an effort by local individuals to show or demonstrate that they are equal to the more affluent visitors. By adopting the symbols and clothing of more affluent visitors, they demonstrate their equality.

Homogenization Homogenization is a secondary effect resulting from the primary effects noted above. Homogenization is basically the move towards the centre or the common denominator for a variety of cultures. By homogenization, we mean that cultures lose the unique features that make them distinctive and idiosyncratic. It is often these features, whether it is the clothing, values, beliefs, traditions, events or symbols, that may make the culture attractive to non-resident visitors. One of the great challenges of tourism development is to maintain a cultural identity important not only for the continuation of the culture itself but also as a resource for visitors, and to do this without commodification and exploitation occurring.

Discussion of tourism impacts

The process of tourism development is similar to other basic industries in the sense that new developments, strategies and policies may lead to anticipated and unanticipated consequences. But for a community to 'export' its tourism product (its natural and cultural heritage in this case) people must come on site to experience and 'consume' it. Whenever non-residents visit an area, impacts, both positive and negative, occur. Some cultures are better suited to adapt and exploit tourism while others are much more sensitive to visitor clothing, symbols and behaviour.

Cultures are not static, but are constantly changing and adapting to new circumstances and environments. But some cultures are more bound by tradition, ritual and expectations of people's behaviour than others. The Sherpa culture, for example, in the upper Khumbu region of Nepal, has adapted well to increased tourism. Such trekking tourism has led to increased wealth while, at the same time, the culture has maintained important religious-based traditions. Other cultures and societies may be more sensitive to tourism or have more difficulty in adapting to or controlling impacts.

Since the effects of tourism may be pervasive and are not always positive, each community has a responsibility to fully explore the potential consequences when developing a tourism strategy. Such a strategy would allow the community to make informed decisions and explicitly deliberate on the trade-offs that may occur with tourism development.

One of the difficulties in developing strategies and making decisions about tourism in communities adjacent to parks is that impacts, both positive and negative, may occur at different levels. For example, decisions to market park-based tourism at a national level lead to increased foreign exchange at that level, but may lead to negative social impacts experienced at the local level. At the local level, tourism-related businesses may benefit from increases in tourism activity, but others may feel increased competition for favourite recreation places. Thus, there may be a dissociation of costs and benefits of tourism development that will need to be resolved through community and park planning processes.

Communities and Tourism: Preparing for the Future

The literature is not unified in how to approach tourism development and address the types of negative impacts identified above. Tourism development may represent structural shifts in a community's economic base, leading to potential corresponding shifts in the distribution of both political and economic power. Such changes can result in conflict, suggesting that processes dealing with conflict should be considered as part of the tourism development strategy.

Processes to deal with tourism development are hampered by several major factors. First, the tourism industry is fragmented, usually consisting of a large number of relatively small businesses in a variety of sectors acting independently. This makes developing the coordinated action needed for a sustainable tourism industry difficult. In addition, park management may have a direct interest in the tourism development strategies that communities implement, suggesting that the development process must be inclusive not only of the stakeholders affected, but also of the relevant park agencies. Secondly, in many capitalistic countries, it is difficult for government, even at the community level, to undertake planning that limits or guides, spatially, temporally and functionally, the type of tourism development the community might desire. This is overlaid, particularly in the western part of North America, with a strong anti-government philosophy coupled with an equally strong conception of private property rights. In these settings, regulating activities on private property for a common goal is often all but impossible.

In other countries, such regulatory mechanisms are well established and accepted. For example, in the national parks of the UK, there are strict regulations and procedures guiding building construction, repair and modification to protect the landscape's cultural integrity. Such regulations are often accompanied by building design guides to make the application and review process less troublesome and more efficient.

Thirdly, tourism development and resulting impacts are often separated in time and space, making projection of impacts and their control difficult. For example, improvements in a local airstrip may reduce limitations on access leading to higher levels of visitation. Such visitation then may increase the demand for lodging and visitor facilities within a park. The increased demand for facilities may eventually require larger sewage and potable water facilities. These effects may occur long after the initial decision for improvements in the airstrip, and may never have been considered in that decision.

Every community, whether located in or adjacent to a national park, or in any other place, needs to plan for tourism. Such planning is needed whether a community decides to pursue tourism actively as an economic development policy or not. In the latter case, planning is needed to prepare the community for the tourism that will occur regardless of the community's attitude towards tourism. In situations where communities have decided to pursue tourism as part of an overall economic development policy, there is a need to develop a comprehensive, adaptable and responsive approach to tourism development.

In this section, we propose certain principles and guidelines that are designed to prepare communities for a future in which tourism is an active component. We emphasize here that tourism is not simply a matter of printing better brochures and building facilities (so that visitors will come). Brochures and facilities may be needed, but they occur among the latter stages of tourism development. Building and implementing a successful tourism development policy requires a thoughtful considered analysis of: (i) the relationship between the community and protected area; (ii) the goals for both the community and the park; and (iii) the perceived efficacy of tourism in attaining these goals. In many respects then, a tourism development strategy requires that a community thinks carefully about its future and develops policies that will assist it in achieving that desired future. However, there are some more specific considerations that would be useful in developing such a policy, as noted below.

Develop the capital necessary for tourism

Capital comes in three forms: social, natural and manufactured. Social capital involves the human skills and abilities needed to make manufactured capital out of natural capital. In the context of this book, natural capital is the biodiversity and natural resources that exist within a park or protected area. Manufactured capital consists of the things and services provided by the natural capital as a result of the application of social capital.

Tourism is very much a knowledge- and information-based economic sector, particularly when involving natural environments and protected areas. The natural capital in parks is 'converted' to manufactured capital through the skilful use of knowledge and information about biodiversity and natural resources, resulting in opportunities to learn, appreciate and enjoy the existing natural capital. The natural capital within parks is preserved during sustainable tourism development, and tourists use it through interpretation and recreational facilities, but in non-consuming ways.

Critical to this process is social capital. Social capital involves the capacity of humans to deal with problems and challenges in the future. More specifically, social capital entails leadership, problem-solving and organizational capabilities, entrepreneurial and managerial talents and proficiencies in analysing data, visualizing the future and perceiving opportunities. Without the presence of social capital – including knowing how to apply it skilfully and sensitively – tourism development may lead to unacceptable social and environmental consequences, ineffective promotional programmes, dissatisfied visitors, misallocation of resources and poor returns on investments. Thus, tourism development is not simply a matter of printing brochures and building lodges, restaurants and visitor centres, but involves a variety of activities acting in concert to reinforce each other.

Strengthening social capital (and thus the capacity to manage tourism) is critical to successful tourism development. This may be done through training in business practices, leadership, planning and political activism; it may include education, both formal and continuing; and it may require increased awareness of park and tourism management principles, concepts and research. In some cases, highly trained individuals may move into the community and, through both formal and informal interpersonal networks, lead to capacity building.

Prepare development plans that are adaptive and responsive to changing conditions

Sustainable tourism planning may be conceived as a process for identifying a desired future and development of a pathway to that particular future. Unfortunately, there is always a great deal of uncertainty concerning the effectiveness of any particular pathway in achieving a desired future. In addition, the overall contextualizing future that may occur is based on certain implicit assumptions (such as political stability or energy prices) that in a few years may prove to be invalid. A tourism plan needs to be adaptable in the sense that the pathways to the future may need to be changed because of shifting circumstances, even if the vision of the desired future remains the same. The plan should be responsive in the sense that needs often change; our assumptions about what is desirable may shift as we gain experience with implementing an economic development strategy.

Thus, the plan should be developed and written in such a way that recognizes the uncertainty that occurs with predicting the future and what it may be like. One of the ways in which this can be done is to ensure that monitoring of critical indicator variables occurs periodically throughout the implementation process. These variables are items that may reflect goals that have been established, issues of concern or substantive areas (such as employment or per capita income) that the plan attempts to address. That plan should contain multiple opportunities for feedback and evaluation of progress towards achieving goals. Strategies for dealing with economic, political or natural 'surprises' – such as unexpected increases in the price of petrol – would be appropriate components of such plans.

Tourism development should be inclusive in character

The purposes behind tourism development are to enhance economic opportunity, maintain or improve the quality of life and protect a community's natural and cultural heritage (Fig. 9.3). Development activities may permeate a community, affecting various groups differently. This requires the tourism planning and development strategy to be inclusive of those that are affected in order to completely identify the values of interest, the positive and negative consequences and how such consequences are distributed through the community. Actions needed to address the needs of those adversely affected require a public participation process that recognizes that negative impacts should be addressed. Such public participation, properly structured, provides opportunities for dialogue, learning,

Fig. 9.3. Nature-based tourism stimulates craft industries. This cooperative store in Kenya provides high-quality carvings to the safari tourists. Craftwork employment is highly desirable and over time leads to enriched community culture. Craft store in Malindi, Kenya. (Photographed by Paul F.J. Eagles.)

ownership in the plan and strengthening relationships among groups within the community. It also allows explicit consideration of effects and how they are distributed.

Tourism development will often occur in contentious, volatile and politically difficult environments. Some residents may not want tourism while others will enthusiastically support it. The process of developing tourism strategies is one that recognizes the legitimacy of different views and types of knowledge, and seeks, as mentioned earlier, to accommodate them.

The science associated with tourism development is anything but definitive and conclusive. Thus, tourism development may be viewed as a messy or wicked problem. In these settings, planning processes must focus on gaining a consensus about what futures and actions are needed within the community and on learning: realizing that we do not necessarily understand the cause–effect relationships that form the basis for policy. This suggests that tourism development policies are necessarily experimental in character and thus the planning process must be adaptive.

Clearly understand the consequences of tourism development

Tourism development, regardless of what national policies may have been set in motion, occurs at the local level. Informed decisions require that the analysis of potential tourism strategies is comprehensive so that decision makers, with input from the affected public, make informed decisions about the nature of the trade-offs that will be made through a tourism strategy. When the level of tourism is increased and economic benefits begin impacting the community, there will also be social and cultural impacts, some of which may be negative. Understanding the consequences of a tourism strategy requires that we depict all the consequences, not just the economic benefits of tourism. This allows the community to make decisions about how much and what kind of tourism development to encourage, how many and what type of tourists to attract, and what will be done with the revenues to the community as a result.

Understanding the trade-offs is critical to informed decisions. It may be that communities are willing to tolerate increased competition – at least to a certain level – for favourite recreation places in order to achieve enhanced economic opportunities. But without a comprehensive analysis of trade-offs, it will be difficult to understand the relationship between the benefits and costs of tourism development. Requiring such a comprehensive analysis also forces the community to move away from the type of uncritical promotion that often accompanies proposals for tourism development. While the uncritical promotion may be useful in generating enthusiasm and organizing community resources, there is the danger that the positive outcomes will be over-sold and the negative consequences will be marginalized.

Inventory places and localities important to the community

In many communities, there are places that contain special meanings and values or are of spiritual significance to members of the community. These are places where it might not be appropriate for tourism development or promotion. It is important, therefore, in terms of mitigating potential negative impacts, that the community identifies these places for itself and commits to avoid promotion of these locations. The community may also adopt rules determining who gains access, when and how.

Understand the relationship between the park and the community

Communities within or immediately adjacent to national parks and protected areas need to take steps to understand the dependency and scale of the linkage, in terms of economics, culture and policy, between the community and the park (see Fig. 9.4). While many of these linkages are economic in character, as we noted earlier, there are also social and political dimensions to them. In particular, communities are often unaware of the management policies and issues confronting a specific park, and may even be unaware of how their tourism promotion strategies affect biophysical and social conditions within the park.

Often, the only relationship that communities have with adjacent parks is adversarial in nature; only the negative consequences to the community of park management policies are known or understood. Communities that have strong social, political and economic ties to a park are among its greatest advocates, providing both positive and negative feedback to the park administration as well as to higher-level civil authorities and legislative bodies.

Develop strategies that enhance positive consequences and mitigate negative effects

Many of the strategies that communities adjacent to or within a protected area may initiate will have consequences not only to the community but to the protected area as well. While some of these consequences may be a surprise, others may be easily predicted. It is important therefore that the consequences are dealt with as part of any tourism development, marketing or promotional strategy. In this sense then, communities need not only acknowledge the potential effects, but also ensure that appropriate monitoring occurs that will be able to portray those effects and suggest potential mitigation strategies. For example, a community promoting

Fig. 9.4. Accommodation is one of the most lucrative sources of income for tourism. Limited accommodation within most parks provides business opportunities for entrepreneurs in local communities outside the park. Springdale is one gateway community to Zion National Park. It provides a full range of services to park visitors. Bed and breakfast sign in Springdale, entrance community to Zion National Park, USA. (Photographed by Paul F.J. Eagles.)

nature-based tourism dependent on a nearby park must understand that increasing levels of tourism may lead to changes in the character, intensity and location of the impacts on the protected area. Through the mutual fact-finding mentioned earlier, community tourism planners and the park administration can identify and pursue the policies and actions needed to mitigate effects.

Secure necessary conditions for plan implementation

Preparing a community for tourism and the future ultimately leads to a formal planning document, which explicitly states desired futures, policies and actions needed to achieve that future, and a strategy to monitor implementation and effectiveness of actions. There are a number of elements to successful community preparation that would characterize such planning. These include: (i) a clearly written planning document that is easily accessible to those with interest; (ii) the public and private understanding, ownership and acceptance of the plan needed to generate funding and any necessary policy or legal revisions; (iii) a competent and trustworthy agency or group charged with the leadership needed to implement the community vision; (iv) sufficient funding and financing capabilities for implementation and monitoring; (v) political stability that ensures continuity in the vision and plan and commitment to implementation; and (vi) strong, yet sensitive, political leadership that supports the community plan. Such leadership would be able to communicate and negotiate with other agencies, such as park organizations that may be affected by the plan.

Monitoring is essential to any tourism implementation strategy. Monitoring is a systematic and periodic measurement of key indicator variables. Monitoring will show whether or not the proposed strategy has been successful in achieving objectives, where corrections may be needed in that strategy and, through evaluation of monitoring data, determine unanticipated consequences. When developing a tourism strategy and considering both the positive and the negative consequences, key indicator variables are identified that will be used in the monitoring phase of the strategy implementation.

Case Study Number 6: Kakum National Park and Conservation Area (Ghana)

A Non-governmental Organization Working with a Park Agency to Develop an Ecotourism Industry in a National Park

Kakum National Park is located 20 km north of the seaside town of Cape Coast in Ghana's Central Region. Kakum National Park and the adjacent Assin Attandaso Resource Reserve cover 350 km^2 of tropical moist forest. In 1932 Kakum was declared a forest reserve and managed for timber extraction.

In 1992, Kakum and Assin Attandaso were redesignated as national parks and jointly managed as the Kakum Conservation Area. The area provides habitats for the globally endangered forest elephants, bongo, yellow-backed duiker and Diana monkey, an estimated 550 butterfly species, 250 species of birds and 100 mammal, reptile and amphibian species. This area is part of the Guinean Forest Region of West Africa, a globally important area of biodiversity.

Recognizing the global importance of the area, Conservation International initiated a programme to develop an ecotourism industry in the national park. The goals of the programme are to provide economic alternatives to logging for the local community, encourage forest conservation and provide a local sense of pride in the natural resources.

Kakum National Park hosts a number of internationally recognized endangered species. However, many first-time visitors to the park are disappointed when the animals prove to be difficult to see. One reason is that many live high above the ground, hiding in the dense tangle of vegetation of the forest canopy. It was important that people were encouraged to visit the national park in order to create an ecotourism industry that could help support the park and the local community. To provide better viewing of the wildlife, an aerial walkway was constructed, the first of its type in Africa. The walkway is 333 m in length and is suspended approximately 27 m off the ground by eight huge emergent trees. Surrounding each support tree is a wooden platform where visitors and researchers can stop and spend time observing the rainforest from heights of up to 12 storeys off the ground. Great care was taken to ensure that the trees, which support the walkway, are protected from injury, therefore nothing is nailed or bolted to these 300–400-year-old trees. The canopy walkway offers students, tourists, researchers and policy makers access to the rainforest canopy.

The canopy walkway was designed and constructed under the guidance of Conservation International's Ghana Program. With financial support provided by the US Agency for International Development and by Conservation International, the park development has been carefully planned to provide maximum sustainable benefits to both wildlife and local communities through the collaboration of Ghanaian and US experts.

On Earth Day 1997, a visitor centre opened in Kakum National Park. It supports educational, social and ceremonial functions, with a special emphasis on demonstrating the unique relationship between nature and culture in West Africa. The interpretive centre also highlights the park's flora and fauna and the importance of protecting Ghana's natural heritage. The walkway, the visitor centre and the park's natural resources have attracted local and international media as well as more than 20,000 visitors to the park, up from 700 in 1993. This rapid and sustained growth in visitation after 1994 was due to high visitor satisfaction from the services and facilities developed by the ecotourism programme.

In 1995, tourism surpassed timber to become the fourth largest generator of foreign currency in Ghana. As the first protected area in Ghana to receive major support for building visitor facilities, Kakum National Park has the potential to set a standard for all future design and development of Ghanaian protected areas.

This project was planned and implemented under the concept that ecotourism can be both an effective conservation tool and a successful community development model. The Kakum project uses a community-based approach to tourism development, which can be part of a successful conservation strategy. An effort was made in every possible way to provide tangible benefits to local people. These included the purchase of agricultural products for the Kakum restaurant, the purchase of furnishings, crafts and services from local artisans, the provision of guide training to local teachers and the creation of full-time, direct and indirect employment. Help was given to local people by assisting them in establishing and managing their own ecotourism businesses. These businesses created jobs that directly depend on a healthy environment and motivate people to protect their surroundings. The hope is that people who earn their living from ecotourism are likely to defend their natural resources against more destructive activities, such as logging or mining.

The success of the Kakum project is due to four critical aspects. First, it was a well-conceived local government initiative. Secondly, USAID provided long-term and significant financial support. Thirdly, there was consistent and high-quality technical direction provided by Conservation International. Fourthly, Ghana had a stable political environment and an expanding economy over the period of project development.

The Conservation International project involving Kakum National Park in Ghana was awarded the 1998 Global Tourism for Tomorrow Award as the best example of environmentally sustainable tourism in the world. It is an excellent example of local community development, carefully structured to create an ecotourism economic alternative to resource exploitation (website: www.conservation.org/web/fieldact/regions/afrireg/ghana.htm).

Conclusion

Protected areas and communities are linked not only by the presence of a tourism industry and the need for its management, but by common interests as well. Identifying and protecting those interests is fundamental to preparing communities and protected areas for the positive and negative consequences of tourism.

Such preparations can neither be separated functionally nor can they be conducted outside venues that do not consider community–park linkages. Such linkages occur over various timeframes and components of the community.

Preparing for the consequences of tourism development, whether positive or negative, can best be seen as requiring a process that is inclusive of all potential interests affected. Any tourism development process in a community that is linked to a protected area is not simply a matter of boosting visitation to the community. It also involves serious discussion not only about the product that the community will offer but also about the effects of such promotion on the nearby park.

Perhaps the single most significant step a local community can take as it begins considering tourism as an economic development alternative is to strengthen its social capital so it has increased its capacity to deal with the changes and challenges that tourism presents. Increasing social capital requires the involvement of a variety of individuals within the community (who play varying roles: business people, guides, planners, activists, supporters, coordinators, decision makers) and this will take time. A few workshops here and there are simply not adequate in most situations. There must be a long-term commitment to capacity building, education and technical assistance from governments and NGOs. Eventually, the community capacity will be boosted, leading to a situation where it decreasingly needs help.

References

Litchfield, C. (2001) Responsible tourism with great apes in Uganda. In: McCool, S.F. and Moisey, R.N. (eds) *Tourism, Recreation and Sustainability: Linking Culture and the Environment.* CAB International, Wallingford, UK, pp. 105–132.

Machlis, G.E. and Field, D.R. (eds) (2000) *National Parks and Rural Development: Practice and Policy in the United States.* Island Press, Covelo, California.

McLaren, D. (1997) *Rethinking Tourism and Ecotravel.* Kumerian Press, West Harford, Connecticut.

Sherpa, M.N. (1993) Grass roots in a Himalayan kingdom. In: Kempf, E. (ed.) *The Law of the Mother: Protecting Indigenous Peoples in Protected Areas.* Sierra Club Books, San Francisco, California, pp. 45–51.

Further Reading

Butt, N. and Price, M.F. (eds) (2000) *Mountain People, Forests, and Trees: Strategies for Balancing Local Management and Outside Interests*, Synthesis of an Electronic Conference of the Mountain Forum 12 April–14 May 1999. The Mountain Institute, Harrisonburg, Virginia.

Rogers, P. and Aitchison, J. (1998) *Towards Sustainable Tourism in the Everest Region of Nepal.* IUCN Nepal, Kathmandu.

CHAPTER 10
Tourism in Marine Protected Areas

Elizabeth A. Halpenny

Introduction

This chapter is dedicated to addressing a special area of study: coastal and marine reserves and how to manage tourism and recreation in these reserves (Fig. 10.1). Coastal and marine environments provide special challenges to tourism and recreation planners; these problems are outlined here followed by examples of how these challenges have been addressed in Tanzania, Netherlands Antilles, Kenya and Indonesia. An outline of various tourism activities in marine protected areas is also provided along with a description of their related impacts.

Coastal Environments and Marine Protected Areas

Much discussion has been devoted to describing what a coastal area is (Hinrichsen, 1998; Kay and Alder, 1999); however, for the purpose of this chapter, a coastal area is defined as 'that part of the land affected by its proximity to the sea and that part of the ocean affected by its proximity to the land . . . an area in which processes depending on the interaction between land and sea are most intense' (Sorensen cited in Hinrichsen, 1998). While tourism activity occurs in numerous areas along the coast and in large lakes and the ocean, the focus of this chapter is on planning and management of tourism in marine protected areas (MPAs). For most resource managers and tourism planners, it is often difficult to envisage a marine protected area, unless they have had extensive experience in working in marine and coastal environments. An MPA is defined by the World Conservation Union (IUCN) as 'any area of intertidal or subtidal terrain, together with its overlying water and associated flora, fauna, historical and cultural features, which has been reserved by law or other effective means to protect part or all of the enclosed environment' (Kelleher and Kenchington, 1991). Examples include protected areas around coral reefs, underwater sea mounts and geothermal vents. MPAs are not just marine based, they also include terrestrial protected areas that contain or border shorelines, estuaries or wetlands; thus their boundaries encompass an oceanic shoreline, thereby providing coastal protection (Silva *et al.* cited in Tisdell and Broadus, 1989). Additionally the aquatic and shoreline environments of large freshwater lakes and rivers can also host MPAs (e.g. Fathom Five National Marine Park and Saguenay Marine Park in Canada).

Unlike their terrestrial counterparts, marine and coastal environments provide additional and different challenges for tourism planners and managers. Several of these differences are outlined below.

The Challenges of Tourism in Terrestrial versus Marine Protected Areas

Over 50% of the planet's human population now lives and works within 200 km of a coast on approximately 10% of the Earth's land area (Hinrichsen, 1996 cited in Hinrichsen, 1998). The number of people moving to coastal areas continues to expand, bringing with it increasing pressures on the resources found in coastal areas, such as water, food and space. Often this development is unplanned, leading to dramatic declines in the health of these coastal ecosystems. Tourism and recreation are two of the main activities that take place along these coasts. In fact the growth rate of marine tourism has exceeded most other forms of tourism. For example, whale watching displayed an average growth rate of 10% in the 1990s compared with an average annual 4.3% increase in world tourist arrivals (Cater and Cater, 2001). Attempting to manage tourism activity in a marine protected area situated along a densely populated coast is especially challenging. To complicate this, most coastal and marine environments are commonly held properties, requiring cooperative agreements on their use. MPA managers must work with terrestrial and marine-based stakeholders including agricultural and fishing interests,

Fig. 10.1. Tropical marine coastal tourism often involves beaches that are very durable and can handle large levels of human traffic. It also involves nearby coral reefs and mangrove ecosystems that are very sensitive and easily damaged by visitor use. Playa Del Carmen, Mexico. (Photographed by Paul F.J. Eagles.)

urban waste-management agencies, forestry, recreationists and other interests.

In the face of development pressure experienced in coastal areas, MPA managers must also deal with the public's weak understanding of marine resources. Because most of the active processes that occur in the ocean and in coastal areas take place underwater, most people have little or no understanding of the importance of marine ecosystems and marine conservation. With low levels of awareness regarding the importance of marine conservation comes weak administrative structures and very low funding for MPAs. Also, when recreation and tourism need to be managed in a MPA, an extensive amount of education of the public must be undertaken to ensure support from local stakeholders. Recently, greater public awareness of marine resources and the importance of marine conservation has been documented in the United States by the marine educational campaign project called Sea Web (Spruill, 2001), which has contributed to an expanded programme of marine reserve development in that country. The expansion of MPA programmes has also appeared in Australia and Canada; however, overall public awareness still lags far behind knowledge levels on the importance of terrestrial parks and resource conservation.

This said, MPAs have a greater potential for educating the public about conservation. Marine tourism and recreation have the potential to create a much stronger and more positive public image for MPAs, similar to what has occurred over time in terrestrial parks. One might say that public attitudes towards marine environments are at an earlier stage, possibly 50 years behind terrestrial environments, at least in developed countries (P. Eagles, Almonte, Ontario, 2001, personal communications).

The greatest challenge facing the MPA manager is dealing with the rapid movement of materials in and out of an MPA ecosystem. Massive flows of materials and organisms drift through MPAs, bringing with them the essential building blocks for life forms within the MPA such as nutrients and substrate. These flows also bring the negative legacy of terrestrial and marine resource development, excessive nutrient levels from waste, solid waste, increased sedimentation levels resulting from terrestrial erosion and coastal dredging, and so on. MPA managers also must take into consideration the rapidly changing marine environment created by currents, storms and tides. These forces can have a tremendous impact over a relatively short timescale in comparison with impacts found in terrestrial parks. The volatility of the land and ocean interface creates many challenges, especially when it comes to developing tourism and recreation infrastructure such as moorings and docks. This volatility also leads to a higher degree of risk for people and property from natural hazards (see Fig. 10.2).

Coastal areas, one of the chief regions that MPAs are designed to protect, are unique and form one of the most productive ecosystems on the planet. In them are located wetlands, coral reefs, sea grass beds and so on. Wetlands such as salt marshes and mangrove swamps are highly productive systems producing more wildlife both in numbers and in variety and more primary plant growth than any other habitat on Earth. They are also highly effective natural filters (of nutrients, pollution and silt) and provide invaluable shoreline stabilization. Mangrove forests

Fig. 10.2. Water-based recreation has risks. This photo shows mid-lake rescue instruction for novice wilderness canoeists. Marine park managers use a wide variety of interpretation, regulation, enforcement and search-and-rescue programmes to handle water recreation risks. Canoe safety demonstration, Algonquin Provincial Park, Canada. (Photographed by Paul F.J. Eagles.)

provide habitats for over 2000 species of fish, shellfish, invertebrates and epiphytic plants. Their importance to humans is illustrated in an example from Fiji where approximately half of all fish and shellfish harvested by commercial and artisanal fishers are dependent on mangrove swamps for at least one stage in their life development. Sea grasses, like wetlands, are also highly fertile ecosystems, with productivity levels comparable to those of agricultural crop lands. They also provide an important habitat for marine species and stabilize coastal areas. Coral reefs yield multiple benefits: they protect coastlines from erosion and storm damage, provide habitat to tens of thousands of species of fish, shellfish and invertebrates, and form an important link in cycling nutrients from the land to the open sea. Nearly a third of all fish species live on coral reefs, leading Salm to state that 'coral reefs are self-perpetuating fish farms which produce high-quality protein from essentially empty sea water' (Hinrichsen, 1998). All of these ecosystems are under extreme pressure and are disappearing at an alarming rate in the face of development. The abstract from Wilkinson's (2000) *Status of the Coral Reefs of the World* illustrates this:

> Coral reefs of the world have continued to decline since the previous GCRMN report in 1998. Assessments to late 2000 are that 27% of the world's reefs have been effectively lost, with the largest single cause being the massive climate-related coral bleaching event of 1998. This destroyed about 16% of the coral reefs of the world in 9 months during the largest El Niño and La Niña climate changes ever recorded. While there is a good chance that many of the 16% of damaged reefs will recover slowly, probably half of these reefs will never adequately recover. These will add to the 11% of the world's reefs already lost due to human impacts such as sediment and nutrient pollution, over-exploitation and mining of sand and rock and development on, and 'reclamation' of, coral reefs.

Additionally the value of these unique and highly productive ecosystems is just beginning to be understood and a strong base of scientific knowledge has been slow to accumulate. Developing strategies for these ecosystems has been slower than in terrestrial areas; as a result, the management of these resources and tourism activity in MPAs makes planning and development especially difficult and often plagued by spectacular failure (e.g. the collapse of the Northeast Atlantic Cod Fishery).

Another interesting difference between tourism in MPAs and terrestrial parks is the type of products and experiences offered to tourists. In general terrestrial parks have many more cultural and historical heritage resources to draw on (Brylske, 2000). MPAs are almost entirely dependent on nature-based activities and attractions. There are some exceptions such as shipwreck diving or visiting coastal archaeological ruins; however, for the most part, MPA managers have a narrower range of options to work with in developing tourism products in an MPA and managing the type and level of visitor activities. MPA managers therefore have a narrower margin for error as so much of the visitor income for surrounding gateway communities and the parks' own entrance and services revenue depend on a handful of nature-based resources.

Marine protected areas face another management issue, directly related to tourists' and recreationists' access to marine reserves. Due to the nature of coastal and marine areas it is impossible to set fixed boundaries for MPAs such as fences; as a result it is often difficult to control entry, restrict access to highly sensitive areas, administer an entrance fee or post educational signs. Terrestrial protected areas do not experience this challenge to the same degree. Box 10.1 summarizes the differences between MPAs and terrestrial protected areas.

Tourist Activities and Their Impacts in MPAs

Ecological impacts of tourism in MPAs

Tourism activities in marine protected areas take many forms. They can range from 'passive' activities such as sunbathing and photography to 'active' forms such as surfing and boating (Kenchington, 1992).

A diversity of boating operations can occur in an MPA including day trip vessel operations to reefs and island destinations, extended charter

Box 10.1. How MPAs differ from terrestrial protected areas.

1. Common property resources which typify most coastal and marine areas require cooperative stakeholder management.
2. High level of development pressure due to exponential growth of human settlement and extractive resource use along coasts. Often direct competition between tourists and local communities for use of the resources develops.
3. Weak understanding by general public and policy makers of marine environments and the importance of their conservation. This results in inadequate funding for and management of MPAs.
4. Currently MPAs, through tourism, may play a more important educational role in exposing the public to the importance of conservation, just as terrestrial parks did during the 20th century.
5. Unique and highly productive ecosystems found in coastal areas. Massive exchange of materials in a coastal system, and influx of pollutants, etc. from outside sources results in management concerns focused on activities and stakeholders outside MPA. Also, marine organisms are impossible to 'keep' in one place (e.g. by a fence), and roam 1000s of kilometres (Aiello and Hunderford, 1996).
6. It is difficult to observe the overall effects of exploitation in a marine environment; the idea that the ocean is limitless in its resources has been disproved with spectacular results (e.g. Northeast Atlantic Cod Fishery collapse) often catching managers unawares.
7. The complex and dynamic nature of coastal and ocean environments make changes difficult to predict. High degree of risk to people and property from natural hazards exists for MPAs and coastal areas.

Continued

Box 10.1. *Continued.*

8. Multiple points of entry and no visible boundaries makes policing and enforcement, collection of fees, and monitoring and education of visitors difficult. Individual boaters often break park rules out of ignorance due to the difficulty in recognizing park boundaries and restricted zones.
9. Relatively low levels of cultural–historical heritage located in MPAs, and the dependency on nature-based tourism attractions and activities.

boat tours (dive live-aboards and fishing charters) and international cruise ships. Large cruise companies may send small portions of their passengers to enjoy a day trip snorkelling in a local MPA, but smaller expedition cruise ship companies can generally anchor right inside MPAs as their ships are smaller, or they may use small, manoeuvrable, inflated boats to transfer guests to an attraction. Transporting supplies and staff by boat is another aspect of tourism-related boat traffic in MPAs. A motorized form of boating, jet skis are a controversial addition to any MPA and their use has been banned or strictly curtailed in many MPAs. Jet skis are often associated with high noise levels and high speed in inappropriate areas of parks, creating wakes that erode shorelines and habitat. Collisions of jet skis and motor boats with slow-moving mammals such as dugong or manatees is one of the leading causes of fatality among these animals. Motorized vehicles can also degrade sea grass beds through propeller damage.

Other ecological impacts associated with boating include pollution, especially from two-stroke motors used to power boats. Noise, fuel consumption and pollution can be reduced through the use of four-stroke engines now available on the market. Anchoring boats is another challenge for boaters. Moorings established by the management agencies of the park should be used at all times to avoid damaging plant and animal life. Anchors from cruise ships in the Caribbean have been reported to have destroyed areas of coral reefs when improperly deployed. For example, in the Cayman Islands it was reported that a cruise ship anchoring for 1 day caused the destruction of 3150 m^2 of previously intact coral reef, the recovery of which would take an estimated 50 years (Smith cited in Viles and Spencer, 1995). In 1997, Belize was successful in charging one boat operator with negligence when a reef was damaged; the tour company had to pay US$75,000 in damages, which went towards the construction of further moorings (J. Gibson, Belize City, 1999, personal communication). If moorings are not available boat operators are expected to anchor on sand or a similarly barren surface.

Bilge water and its disposal is another item of concern for MPA managers. Bilge water collected in another port can transport disease or alien species to the MPA if it is dumped in the park. Boat operators are encouraged to dump the water at sea or in a pump-out facility. Discharge of vessel sewage and waste is a related disposal concern. No dumping can take place near shore; all boats must comply with the International

Convention for the Prevention of Pollution from Ships at Sea (MARPOL 73/78). Again, pump-out and landfill facilities should be used when available and, if not, then onboard treatment equipment is required or dumping of organic waste is permitted out at sea (but generally not within the boundaries of the MPA).

Non-motorized boating is also common in MPAs and includes sea kayaking and canoeing. These recreational vehicles provide the opportunity for exercise and excellent wildlife viewing platforms. These boats have few impacts on the natural environment; however, some research does indicate that the presence of tourists on these types of crafts may be more disruptive to certain wildlife compared with the presence of motorized vehicles, in part because they can enter areas that motorized vehicles cannot.

Additional active tourism activities include fishing, hiking, windsurfing and surfing. Recreational fishing is generally 'catch and release' in MPAs, if it is permitted at all. Over time, consumptive fishing, such as spear fishing, has declined considerably. As with fishing outside the reserve, over-fishing will result in a decline of sport fish species. No-take reserves, which feature a total ban of fishing in the MPAs, are becoming very popular in the management of fish stocks outside reserves. The impacts of hiking in MPAs are no different from those in terrestrial parks; adherence to 'leave no trace' standards should be encouraged by managers (NOLS, 1998). One of the most common problems associated with hiking (and wildlife watching) in coastal areas is the disturbance of nesting birds along the coast, as their nesting habitat is often a very narrow band along the seashore. Windsurfing and surfing, like sea kayaking and canoeing, have little impact on the natural environment. However their presence in MPAs in recent years in the United States has provided a powerful advocacy voice for marine conservation (Surfrider Foundation, 2001).

Another tourism activity which might be permitted in a marine protected area is 'collecting': harvesting materials for food or beachcombing for souvenirs. Harvesting occurs in some MPAs, especially biosphere reserves where multiple uses are encouraged and traditional uses are incorporated into management schemes. Most MPAs prohibit such activities, especially the collection of souvenirs by visitors, in an attempt to maintain the 'naturalness' of the park and the ecological processes occurring there.

Scuba-diving and snorkelling are the two tourism activities most associated with MPAs. Scuba-diving continues to grow at a very high rate in South-east Asia and Europe; however, growth has slowed somewhat in North America. In the larger scheme of marine conservation, divers have little negative impact on the marine environments they are visiting. However, there have been incidents of divers touching delicate coral or taking pieces of historic shipwrecks, irrevocably damaging the environment they came to visit. Education can play a large role in reducing these impacts. For example, in a 3-year study in Egypt's Ras Mohammed National Park, it was demonstrated that by implementing a single environmental briefing on the delicate nature of the ecosystem and emphasizing low-impact diving techniques, diver contacts with living coral were reduced from 8.3 to 1.5 incidents per dive. Damaging impacts (i.e. contacts where structural damage was inflicted on the corals) were reduced from 6.7, about 80% of the total, to 0.3 impacts, about 20% of the total, per dive (Medio, 1996, 1997 cited in Brylske, 2000). Fin kicks that stir sediment, dangling equipment and touching (e.g. to steady the diver during an underwater photography session) are the most common incidents. Snorkellers are perhaps more prone to injuring reefs as they are generally less experienced in the water and tend to feel uncomfortable in an unfamiliar mask and snorkel; they invariably stop to adjust equipment, often stepping on the reef for support.

One of the greatest pastimes for snorkellers and divers is wildlife watching. Luring fish from their lairs is a tradition for some dive operators who wish to give their clients 'a great show'. Other operators regularly use chumming, using fish parts and blood to lure sharks to a specific area where divers can observe sharks from diving cages. Fish feeding can have profound negative impacts; these include disruption of natural ecological processes within an ecosystem, degradation of fish health due to inappropriate feed, and human endangerment because of increased aggression by fish searching for feed. The practice of chumming was banned in Monterey National Marine Sanctuary in the United States in the late 1990s after extensive protests from local surfers who feared for their lives. However, it is still common in other areas of the world including the Bahamas. Other MPAs do still allow fish feeding; the Great Barrier Reef Marine Park Authority allows fish feeding under very controlled circumstances, using a permit system to designate which operators can feed, what they feed and when.

Another form of wildlife watching in MPAs is by video display of underwater or remote coastal scenes. Tourists can watch what is going on in these remote locations without getting wet. This form of virtual tourism is becoming more popular and has tremendous educational potential with few, if any, negative ecological impacts.

Fish are not the only attraction to be viewed by wildlife watchers in MPAs; others include whales, dolphins, birds, invertebrates in tidal pools, turtles and so on. There are extensive guidelines for watching each type of animal; however, these must be tailored for the specific species and location in which that animal is being viewed. For example humpback whales behave differently from beluga whales and therefore boat operators need different operating standards when taking tourists to see these two very different animals. Also, humpbacks act differently at different latitudes, they may be feeding when they migrate to polar regions but may be very protective of their young when at tropical latitudes. The presence of tourism in coastal areas invariably changes the behaviour of these animals, affecting feeding, breeding and sleeping patterns. The challenge for MPA managers is to minimize the negative impacts. MPA managers need to work with operators and wildlife specialists to determine the most appropriate codes of practice for their region.

Invariably, with wildlife watching comes the need for interpretation and environmental education. These two activities are very important in MPAs. It is through education and interpretation that tourists will understand more about marine environments and become marine conservation advocates. Aquariums, glass-bottom boat tours, video cameras, guided nature walks, self-guided trails, underwater trails, information kiosks, skits and plays, and participation of tourists in research activities are just some of the many tools that can be used to help tourists to understand the environment they are visiting. Education also helps to reduce tourist impacts on the environment in the MPA. These activities are essential in 'marine ecotourism' experiences offered by the MPA or ecotourism operators who use the park (Fig. 10.3).

Another aspect related to tourism in MPAs is the transport network that helps to deliver tourists to and within the park. Roads and car parks along coastal areas can have profound impacts on the health of coastal ecosystems. Bridges, berms, culverts, ditches and impermeable surfaces all associated with road construction can modify water flow patterns, destroy wetlands, disrupt reefs through siltation and so on. The impacts are well documented in Bell *et al.* (1989), Ryan (1991), Molina and Rubinoff (1998) and Sweeting *et al.* (1999). Boats are a second means of

Fig. 10.3. Coral reef ecosystems are highly valued ecotourism destinations. This ecotour, lead by a member of the Heron Island Lodge staff, introduces marine ecology to an Elderhostel group from the USA and Canada. It is through education and interpretation that tourists understand more about marine environments and become marine conservation advocates. Guided reef tour on Heron Island, Great Barrier Reef Marine Park, Australia. (Photographed by Paul F.J. Eagles.)

transport already discussed; however, the facilities used to service boats such as pontoons, docks, marinas, pump-out facilities and so on can also have negative impacts on the ecology of an MPA. For example, research in Australia's Great Barrier Reef showed that pontoons used as platforms for tourists visiting a more remote part of the barrier reef had little impact because of their construction and location, including minimal detriment to marine life due to shading caused by the platform. It was clear that early pontoons had some major impacts on reef biota, but with changes in mooring technology and mooring pontoons over sediment (e.g. sand) rather than reefs, these impacts have been minimized. However, the reefs around the pontoons have shown a small amount of degradation due to the extensive numbers of snorkellers who visit the sites every day (Nelson and Mapstone, 1997). Air travel is a third form of transport. Of particular importance is the use of small aircraft such as seaplanes and helicopters to reach wilderness settings. Negative impacts associated with the use of aircraft to reach wildlife watching sites are increasingly documented. For example, in Antarctica, studies showed that when moving between breeding sites and the sea, penguins reacted to and fled from aircraft at heights of over 300 m and at distances greater than 1 km. At the breeding colony, birds generally sat still, even when helicopters landed at a distance of about 30 m. Telemetry work indicated, however, that the heart rate of birds approached closely was four times the resting rate in the colony in the absence of disturbance. These findings indicate that there is a significant stress caused by the approach of aircraft, even if birds do not flee. Nevertheless, as with all types of tourism activity, not all wildlife react in the same way to the presence of aircraft. Responses generally depend on five factors: (i) the species; (ii) the location; (iii) the history of exposure to disturbance of the populations involved; (iv) the aircraft type; and (v) aircraft activities (WBM Oceanic Australia and Claridge, 1997). Infrastructure related to air travel can also have a negative impact on MPAs and coastal areas. For example on the island of Negros in the Philippines the extension of an airport runway to accommodate larger jets affected the flow of sediments along the coast and eroded the beach sand away from the tourism coast (Hinrichsen, 1998). In short, tourism infrastructure and transport mechanisms can have a profound impact on coastal areas and should receive considerable attention from MPA managers, even if the development is occurring outside the MPA (Fig. 10.4).

The same can be said for tourism accommodations, both within and adjacent to MPAs. In general, accommodation facilities take three forms: camping, lodges and hotels. All three can have a great many negative impacts on ecological resources within the MPA. This can be due to mismanagement of sewage and solid waste, overuse of underground water aquifers, inappropriate placement of buildings and lighting (i.e. built too close to the coast, destroying mangroves or coastal dunes) and so on. Impacts can be very profound, ranging from the destruction of turtle hatching beaches (light pollution disorients the turtles and buildings may

Fig. 10.4. Marine parks require special equipment for access. This large catamaran takes visitors 72 km from the mainland to Heron Island. Tourism infrastructure and transport mechanisms can have a profound impact on coastal areas and receive considerable attention from managers. Heron Island Resort ferry, Great Barrier Reef Marine Park, Australia. (Photographed by Paul F.J. Eagles.)

be built too close to the water on top of the egg-laying grounds) to the degradation of coral reefs (through eutrophication owing to increased nutrient levels from untreated sewage or siltation from construction or the removal of sea grass beds and mangroves). Numerous publications discuss ways to decrease the impact of tourism accommodation and infrastructure on coastal environments. These include Amaral and Lee (1994), Hawkins *et al.* (1995), Witherington and Martin (1996), Sweeting *et al.* (1999), Halpenny (2002) and Mehta and Baez (2002).

Socio-economic impacts of tourism in MPAs

The social impacts of tourism in an MPA can be examined for two groups: local residents within the park and in nearby gateway communities, and tourists. Tourism effects include the impact of other tourists' presence on the tourist's own experience while visiting the MPA. Were they displaced or annoyed through disturbance or crowding? Was the tourist's desire for a peaceful or stimulating recreational experience unfulfilled because of too many visitors or unacceptable behaviours?

Local residents can experience the same displacement and disruption because of too many visitors. They can also experience increased crime and related negative impacts associated with the increased presence of tourism. However, there can be many positive impacts as well for local communities. Tourism in an MPA can create the opportunity for increased income, jobs and business development, although this is not always the case. In a study by Goodwin *et al.* (1997) of Komodo National Park in Indonesia it was estimated that at least 50% of tourist expenditure in the local economy surrounding Komodo National Park leaked out. This was due to the fact that most of the operators who brought tourists to the park used their own boats, or boats chartered by an outside operator, and purchased their goods (e.g. drinking water) from outside sources. In addition, of the estimated US$5–6 million spent by visitors to the region in 1995/96, only US$1.1 million is estimated to have been spent locally, approximately 20% of the total expenditure by tourists, with 80% never reaching the local economy (Walpole and Goodwin, 2000). Nevertheless, positive economic impacts can be substantial. At the local level tourism can contribute significant income to local communities and marine conservation efforts. Hoagland and Meeks (1997, as cited in Hoyt, 2000) found that in the US state of Massachusetts, 150 full-time jobs and 600 part-time jobs were associated with whale watching. In 2000, 1 million whale watchers produced US$24,000,000 in ticket sales (Hoyt, 2000).

The scientific programmes of several research organizations, such as the Center for Coastal Studies and the Cetacean Research Unit, both in Massachusetts, have benefited from the payment of naturalists from the institutes for the performance of interpretation during whale-watching trips. Researchers were also permitted to carry out photo-identification and other research on whale-watching cruises. The value of having a whale-watching boat as a platform for research is estimated at US$1000 a day (Hoyt, 2000).

At the national level marine tourism and tourism in MPAs can also make a significant contribution to the economy. In Australia, an extensive survey found that marine and coastal tourism was a significant proportion of tourism. It represented 50% of international visits and 42% of domestic visits and resulted in an estimated economic value of Aus$22,892 million in 1995–1996 or 2.9% of all Australian expenditure on goods and services. In the same year estimates were made of the value of tourism in the Great Barrier Reef at Aus$1.06 to 1.2 billion (Australian Economic Group, 1998; Office of National Tourism, *c.* 1999). Much of this expenditure spins off into locally-owned businesses. The Great Barrier Reef Marine Park Authority also benefits economically through an Environmental Management Charge which is currently the equivalent of Aus$4 (US$2) per visitor per day for standard tourist operations such as day trips and charters (Wachenfeld *et al.*, 1999; Bruce Kingston, Main Brook, 2001, personal communication).

This highlights the other benefit that tourism's presence in MPAs can bring, which is revenue from visitor entrance and service fees, and related expenditures. As mentioned earlier, MPAs are funded at even lower levels than their terrestrial counterparts. This has lead to an abundance of 'paper parks' and parks that have virtually no management, enforcement or research activity. As a solution, a handful of MPAs in the 1990s designed methods of charging visitor fees to help pay for management expenses. Parks in the Netherlands Antilles, especially Bonaire and Saba, as well as Belize's Hol Chan Marine Park, were some of the pioneers in this effort. Recent research conducted by Lindberg and Halpenny (2001) of tourism-related revenue regeneration strategies for MPAs found that MPAs generally charge US$1–5 per day or $10–20 per year. Visitors are willing to pay this and more if the money is certain to be retained for management efforts at the park and is not sent to a central government treasury. The most successful models of MPA revenue generation now follow this model. A small sampling of the various fees charged by MPAs is listed in Box 10.2.

Revenue generated by tourists' presence in MPAs has been sufficient in some cases to fund the core administrative costs and programmes of MPAs. Ras Mohammed Marine Park in Egypt, Nelson's Dockyard National Park in Antigua (Van't Hof, 1996; Anon., 2001; A.J. de Grissac, Almonte, Ontario, 2001, personal communication) and Bonaire Marine

Box 10.2. Selected MPA visitor fees (in US$) (Lindberg and Halpenny, 2001).

Australia
- Great Barrier Reef: $2 per day
- Ningaloo Marine Park: $7.50 per day (dive fee)

Bahamas
- Exuma Land and Sea Park: $5 per day (private vessels); $1 per foot per day for charter vessels; no charge for Bahamian vessels

Belize
- Hol Chan Marine Reserve: $2.50 per day
- Half Moon Caye: $5 per day
- No charge for Belizeans

Brazil
- Abrolhos Marine National Park and Fernando de Noronha Marine Park: $4.25 per day

Canada
- Fathom Five National Marine Park and Flowerpot Island: $1.90 per day + annual dive fee of $5.20

Egypt
- Ras Mohammed Marine Park: $5 per day for foreigners and $1.20 for Egyptians
- Red Sea Marine Park: $2 per day (dive), increased at end of 2001 to $5 per day

Indonesia
- Bunaken Marine Park: $8 per year for foreigners (dive) and $0.30 per day for Indonesians

Italy
- Miramare Marine Reserve: $2.20 per day + fees for services, e.g. dive trip is $22, snorkelling is $11

Micronesia
- The island of Truk, formerly known as Chuuk: $30 dive tax and $31.50 per week cruising tax for live-aboards

Netherlands Antilles
- Bonaire: $10 per year (dive fee same for locals and foreigners)
- Saba: $3 per dive and $3 per week for snorkellers; residents are not charged
- Eustatius: $12 per year (dive fee) and $12 per night for yachts

Philippines
- Tubbataha Marine Reserve: $50
- Gilutungan Marine Sanctuary: $1 per day for foreigners and $0.50 for Filipinos

Park in the Netherlands Antilles (C. Glendinning, Almonte, Ontario, 2001, personal communication) are three examples of this, chiefly through visitor fees, yacht servicing and souvenir sales. However in other MPAs the revenue still falls short. For example, in research on tourism in three protected areas based in developing countries, Goodwin *et al.* (1997) found that when tourism fees were required to pay for just the provision of tourism services in the park (e.g. basic tourism infrastructure), only one of the three parks made enough through tourism fees to address these costs. When tourism revenue was calculated to cover the costs of all operational expenditures (research, enforcement, tourism, etc.) in the parks, none of the parks was able to adequately address these costs through tourism-related revenue.

Cultural impacts of tourism in MPAs

Tourism's presence in an MPA can affect the culture of local communities. It can also impact on the traditional and historical values of an area. It can displace recreational users and traditional hunters. Tourists, as described earlier, can also damage, intentionally or unintentionally, cultural heritage such as shipwrecks and archaeological sites. Kealakekua Bay on the west side of Hawaii is a good example of this. A popular destination with local recreationists and foreign visitors alike, this bay is famous both as the death site of Captain James Cook and for the opportunity to swim with dolphins. Many users are present at the site and conflicts have been frequent in recent years. One of the most contentious issues is the site where the Captain Cook monument stands, a frequently visited and often mistreated site, which is also held sacred by native Hawaiians.

Tourism in MPAs also has positive impacts; these include increased awareness of cultural sites and the importance of their protection. It also provides the opportunity for cross-cultural exchange of ideas between visitors and local populations, although the latter phenomenon can also bring negative impacts.

In summary, the greatest threats to MPAs come from development and extraction pressures found along the coast. This includes industrial, commercial and residential activities. However, tourism also affects the health of coastal and marine ecosystems, especially through infrastructure development and waste-management practices. Tourism also brings positive impacts including the support of conservation programmes through visitor fees, donations to and support of conservation programmes by tourism operators, employment and entrepreneurial opportunities for local community members, the prospect for cultural exchange, increased awareness of the public regarding the importance of marine conservation, and so on.

In the face of all of these impacts and management challenges MPA managers must find tools to address the problems and opportunities associated with tourism in MPAs. They use tools also used in terrestrial parks such as Limits of Acceptable Change (LAC), recreation opportunity spectrum (ROS) and environmental impact assessment (EIA), tourism optimization management model (TOMM), adaptive management and ecosystem management. They also use marine-based tools such as integrative coastal zone management (ICZM).

Case studies describe how planners and managers address these challenges associated with managing tourism in an MPA. The Chumbe Island Coral Park case study illustrates how a private MPA achieves conservation and community education goals through tourism. The Saba Marine Park case study examines management and planning of tourism activities through the use of LAC. The Watamu Marine National Park case study illustrates conflict resolution and community participation in tourism activities in an MPA. The Bunaken National Park case study highlights how one MPA has tackled the difficult issue of park funding.

Case Studies of Tourism in MPAs

Chumbe Island Coral Park and Environmental Education Centre, Tanzania: private sector supports marine conservation through tourism

Chumbe Island Coral Park and Environmental Education Centre is a private nature reserve that was developed by a company in 1992 for the purpose of protecting a 20 ha island and surrounding coral reefs, 6 miles southwest of Zanzibar Town. The island was gazetted by the Government of Zanzibar in 1994, becoming the first marine park in Tanzania. The island and surrounding area is a rare example of a pristine coral island ecosystem in an otherwise heavily over-exploited area. The reef supports 370 species of fish and over 200 species of sleractinian corals, 90% of all recorded in the region.

The company which operates the park, Chumbe Island Coral Park Ltd (CHICOP), offers day visits and overnight accommodations for up to 14 guests in seven ecobungalows, all built with state-of-the-art ecotechnology (solar water heating, solar photovoltaic electricity, rainwater catchment, greywater filtration and compost toilets) and eco-architectural design. Five former fishers from adjacent villages have been trained as park rangers and now take full responsibility for protecting the area, producing weekly reports on the health of the sanctuary and guiding tourists including divers.

The management of the site by CHICOP is assisted by an advisory committee with representatives of neighbouring fishing villages, the Institute of Marine Sciences (IMS) of the University of Dar es Salaam and government officials of the Departments of Environment, Fisheries and Forestry. The committee meets one or more times per year.

Revenue generated from tourism subsidizes the conservation and education programmes run in the park. Research programmes include joint projects between park staff, volunteers, and national and international universities and conservation agencies. For example, CHICOP is current conducting research on the world's largest land-living crab, the coconut crab, a resident of Chumbe Island. Education projects address both tourists' and local communities' needs. A converted lightkeeper's house now acts as a visitors' centre where visitor orientation and education take place. Self-guided trails have been developed, complete with 'floating underwater information modules' for the underwater trails, and laminated information cards for the intertidal walking trails depicting fishes, invertebrates and molloscs found on the coast. During low season, excursions are provided free of charge to local school children. Many come from schools within fishing communities where the children benefit from learning about marine resources on which their families' livelihoods depend. Schools in Tanzania have no direct environmental education and often Chumbe is the only insight these children have into environmental awareness.

Approximately two-thirds of the US$1 million start-up cost was financed privately by a benefactor, while the rest was covered through grants from various international NGOs and foreign donor organizations. Now CHICOP receives no additional donor support and depends entirely on income from tourism. They are able to come very close to covering the annual budget of US$120,000; however riots in Zanzibar in January 2001 placed severe financial challenges on the company owing to cancelled bookings.

The project, which took a decade to develop, has identified eight key factors which contributed to the success of the project.

1. The involvement of local people in all aspects of the development.

2. The participation of local residents acting as park wardens, who became highly effective in minimizing activities destructive to the reef ecosystem.

3. The careful design of the tourism facilities, resulting in minimal negative environmental impact during construction and operation.

4. The restoration of the native forest and the recovery of the breeding bird populations was considerably enhanced by the complete removal of a population of introduced rats.

5. The protection of the globally significant coral reef was considerably assisted by the tourism project through increased awareness, etc.

6. The gazetting of the marine reserve, the first in the country, by the national government was stimulated by the project and contributes to the continued viability of conservation efforts.

7. The creation of national law for the private management of conservation areas was stimulated by the project and helps protect conservation progress.

8. Chumbe Island now visibly represents part of the Zanzibari and Tanzanian heritage.

The Chumbe Island Coral Park won the 1999 British Airways Tourism for Tomorrow Award as the best example of sustainable tourism in the world (CHICOP, 2000, 2001; Anon., 2001).

Saba Marine Park, Netherlands Antilles: managing tourism in MPAs – an application of LAC

Saba Marine Park, established in 1987, surrounds the island of Saba in the Netherlands Antilles in the eastern Caribbean. The Netherlands Antilles government established the park to protect the unique and high-quality coral formations and fisheries adjacent to Saba. In 1998 an effort was made to revisit the management plan of Saba Marine Park, to address the increased use of park resources including diving and sport fishing. The chief objective was to provide a framework for long-term decisions affecting the park and those who use it. A Limits of Acceptable Change (LAC) approach was used, seeking answers to the following questions: 'given recreational use, what are the acceptable biophysical and social conditions in a protected area? And what are the appropriate and effective actions to maintain those conditions?' In short, the scientists, planners and community members involved in this process were attempting to address the underlying concern of recreational carrying capacity. While elements of the LAC process have been used in other MPAs, this was the first application of the entire [LAC] process to a marine park.

The process first identified values associated with Saba Marine Park. The park was established primarily to protect outstanding values of the reef's systems and the associated fisheries surrounding the island of Saba. However, more specific values were identified through the process including: (i) the marine environment itself (e.g. water quality, fish habitat); (ii) the recreational values dependent on that environment (e.g. revenues derived from dive tourism); (iii) the park and its management organization (e.g. the park enables education, protection, facilities such as a hyperbaric chamber for diver medical emergencies); and (iv) related social conditions (active involvement of community members in the management of the park, traditional activities have continued with the creation of the park).

Goals and related issues were identified for each value recognized by the stakeholders involved in the process. The issues were especially important in that they were the problems that the LAC process was designed to address. Examples of these included the need for improved harbour and mooring facilities, addressing the danger of oil shipping routes located nearby, and the need for the park to increase and stabilize its operating budget.

While fishing and yachting are popular activities in Saba Marine Park, diving is the chief tourism activity. The park generates funds through fees, donations, souvenir sales, grants and the operation of the hyperbaric chamber. Most of the park's budget is generated through diving fees (US$3 per dive). Anchoring is only permitted where there is a sandy bottom and an effective mooring system is maintained. The park is currently experimenting with a temporary 'guide only policy', which requires all visiting divers to dive with an approved guide. This policy is intended to prevent resource impacts due to improper behaviour, as well as simplify enforcement efforts. The park keeps records of how many dives per day are taken, with 2500 recorded in 1988 rising to over 8000 in 1998.

One outcome of the LAC planning process was the development of a new management zoning system for the park. Instead of a zoning system based on activities (e.g. diving vs. yacht anchorages), the planning process based the new zoning system on acceptable conditions. The new zoning system is not intended to replace the activity zones which were previously used in the park, but rather 'prescriptive management zones' will provide a way for the park to more effectively manage resource and social conditions allowed in each zone.

Indicators were identified to monitor the changes to conditions within Saba Marine Park. These include: (i) water quality (turbidity, salinity, temperature and pollutants); (ii) sedimentation; (iii) fish stocks (diversity and abundance); (iv) damage to corals (breakage and abrasion); (v) number of boats (crowding); and (vi) group size (smaller dive groups have a greater opportunity to interact with nature). Standards were set for each of these indicators. For example, sedimentation level standards are set at 10 mg cm^{-2} day^{-1} for resuspended matter and 10 mg l^{-1} for suspended matter. This standard may become more restrictive based on monitoring results. Another example can be seen in the standards for dive group size: in Zone 1 (the most pristine of the zones) 90% of the time group size will not exceed 20 divers, including guides, in Zones 2–4, 75% of the time group size will not exceed 20 divers, including guides.

Preventive and corrective actions were proposed for the real and hypothetical changes in the conditions of the selected indicators. Examples include education, group meetings, fines and the use of permits. The key to making the results of the LAC process a reality is a monitoring programme that works. The monitoring plan, designed to measure changes to all six indicators, was developed to 'provide the best information possible to park managers while being economically feasible'. Examples include the use of sediment traps, a tool pioneered in St Lucia, placed in four locations to monitor sedimentation rates, and the use of daily records from the five most popular dive sites to measure dive group size, with the most intensive monitoring occurring in three of the busiest months (Schultz *et al.*, 1999).

Watamu Marine National Park, Kenya: resolving conflict between boat operators, tourism companies and government managers

The Watamu Marine Park, located on the Indian Ocean, was established in 1967 as one of Kenya's first marine parks. By the 1990s, increased visitor numbers resulted in competition on the beaches for tourists wishing to travel to the fish-viewing areas of the marine park. Local boat operators vied with hotels for the business of transferring clients. The hotels, in turn, were not supportive of the local operators for legitimate as well as perceived reasons. These focused on issues of marine safety and liability insurance. The local community was increasingly incensed that it had no say in the management of the park, received no benefits or income from it, and could not use it without paying – yet they had been born there. 'Very real threats were made against the park, the park authorities and visitors' (N. Inandar, Almonte, Ontario, 2000, personal communication).

Working with the Kenya Wildlife Service, and funded through a USAID programme, meetings were held with the local boat operators to establish ways of helping them to help themselves, and to tackle the concerns of the local hoteliers. From these meetings, the boat operators were organized to form a group called the Watamu Association of Boat Operators (WABO).

USAID provided some funds to secure life jackets as well as resources for the maintenance of the boats, paint and other materials. The Wildlife Service agreed to help provide a small office, through funding from the Dutch government, as well as training programmes on marine tourism practices. Finally, the group was able to secure insurance for themselves and for their passengers.

Meetings were held with local hoteliers to encourage them to accept the changes and to encourage them to negotiate in good faith with WABO, in order to subcontract the boats from WABO. While the practical concerns had been resolved, there were several management issues, especially to do with guiding quality and customer satisfaction that had to be resolved.

The end result, achieved over 3 years, was that some of the hotels agreed to subcontract this aspect of their business to WABO. Hotels make contracts with the group for the guiding and conveying of their tourists to the respective areas of the marine parks. Instead of fighting on the beach for the lone tourist, WABO negotiates by the boatload with the hotels!

As the group is now involved in the daily activities of the protected areas and is earning an income from the park, there has been a noticeable change in attitude towards the park. The members are more supportive of conservation efforts in the park and are more willing to assist with the management of the park as they are receiving an income directly from the park (N. Inandar, Almonte, Ontario, 2000, personal communication).

Bunaken National Park, Indonesia: developing financial stability for an MPA through collaboration

Bunaken National Park, established in 1991 on the northern tip of the Indonesian province of Sulawesi, has rich biodiversity, including extensive mangrove forests and coral reefs. For years it suffered from a lack of funding resulting in weak management and enforcement of protection laws; dynamite and cyanide fishing threatened reefs and illegal forestry endangered mangroves. Several groups have worked together to establish a fee for visitors to the park. Local dive operators were very supportive of the initiative, they were involved from the inception of the project, working with park managers, international conservation agencies and Indonesia-based NGOs.

There are three general groups of visitors, divers, backpackers and local day visitors. A willingness-to-pay survey determined that visitors would pay an entrance fee of at least US$12.50. However, the sample for the survey was made up largely of backpackers, a more budget-conscious group, and it is speculated that if the survey sample had focused more on the 10,000 dive tourists who visit each year the result would have been higher, perhaps US$20.

For the majority of respondents to the willingness-to-pay survey their chief concern was the management of the collected fee. Visitors wanted to see the revenue go towards conservation programmes in the park, rather than into the coffers of the government or the pockets of local officials. To address this issue, a pilot project was proposed for Bunaken, and the government was lobbied for the creation of a more decentralized approach to fees management. The 'dive industry was a key ally in lobbying the government to pass the law' that would change how the fee revenue would be distributed (M. Erdmann, Almonte, Ontario, 2001, personal communication). The Bunaken National Park Management Advisory Board (a multi-stakeholder board consisting of representatives from the dive industry, environmental NGOs, academia, villagers from within the park and government officials) was created, which receives 80% of the fee revenue, while 20% is split between national, provincial and two district governments.

The fee was developed and initiated over a 10-month period, and came into effect in March 2001. Indonesian visitors pay a fee of Rp2500 (US$0.30) and foreign visitors (divers, snorkellers, backpackers) pay Rp75,000 (US$8). Residents within the park are exempted. The managers and Board chose to introduce a relatively low fee for the first year for several reasons: (i) to minimize industry and especially backpacker opposition; (ii) to prevent government officials from 'eyeing' the funds collected as a treasure trove to delve into; and (iii) to 'prove' to tourists that their fees are really doing something before asking for a larger fee. By starting small, they could avoid overly high expectations from tourists. The managers and Board estimate that it will require approximately

US$250,000 per year at a minimum to manage the park; given current estimates of approximately 10,000 visitors this would mean an eventual fee increase to US$25 year^{-1}. The system is based on Bonaire Marine Park's model, in that when a visitor pays his or her fee at one of two entrance gates within the park, or to a dive operator or travel agent (who buy passes in bulk from the Bunaken National Park Management Advisory Board) he or she receives a waterproof entrance tag that must be worn. As in Bonaire, the tag has become a collector's item. Indonesian day visitors receive paper tickets, as with other national parks.

The implementation of the fees has gone very well. Diver and dive operators are very supportive. Some opposition has been expressed by travel agents who sell a small number of tours to the park. Their chief concern was that they were not consulted from the beginning and were not informed about the fee before their rate lists were published for 2001, thus they could not adjust their prices accordingly. Travel agents are now actively involved in the process, helping the Board to make decisions about how to spend the revenue. The other group that remains in opposition is price-conscious backpackers. Despite an active campaign to inform travellers about the need for the fee and how it will be used for conservation within the park, backpackers remain unsupportive.

Another group whose involvement is increasingly sought is local villagers. Once the fee programme was launched they became more and more 'suspicious' of where all the money was going. An extensive 'socialization' campaign was implemented to let locals know just how the revenue was being used. Other educational campaigns include the development of a frequently asked questions sheet about the entrance fee and where the fees are going; press releases and packages to numerous local newspapers, travel guides (e.g. *Lonely Planet*) and Asia-based diver and travel magazines; an announcement was sent out to all the wholesale dive operators worldwide who take tours to Bunaken; and large neon signs were placed in the arrival halls of the airport (Lindberg and Halpenny, 2001; M. Erdmann, Almonte, Ontario, 2001, personal communication).

References

Aiello, R. and Hunderford, N. (1996) Successes and Failures of Global Reef Management. Reef Tourism 2005 Project. A report prepared for the Marine Tourism Industry of the Cairns & Pt Douglas areas of North Queensland, Australia.

Amaral, M. and Lee, V. (1994) *Environmental Guide for Marinas: Controlling Nonpoint Source and Storm Water Pollution in Rhode Island.* Rhode Island Sea Grant and Costal Resources Center, Narragansett, Rhode Island.

Anon. (2001) Creating self-financing mechanisms for MPAs: three cases. *MPA News – International News and Analysis on Marine Protected Areas* 2(8), 1–3.

Australian Economic Group (1998) *Measuring the Economic Input of Coastal and Marine Tourism.* Executive Summary. Department of Industry, Science

and Tourism (Australia), Brisbane, pp. i–viii.
Bell, P.F.R., Greenfield, P.F., Hawker, D. and Connell, D. (1989) The impact of waste discharges on coral reef region. *Water Science Technology* 21(1), 121–130.
Brylske, A.F. (2000) A Model for Training Tourism Professionals in Tropical Marine Resources Management. Unpublished paper, Instructional Technologies, Inc., Cape Coral, Florida, 18pp.
Cater, C. and Cater, E. (2001) Marine environments. In: Weaver, D.B. (ed.) *The Encyclopedia of Ecotourism*. CAB International, Wallingford, UK, pp. 265–282.
CHICOP (2000) The Chumbe Island Project – the first private marine park in the world. Chumbe Island Coral Park Ltd. Press release.
CHICOP (2001) *Chumbe Island Coral Park*. Homepage of the Chumbe Island Coral Park Limited, online. Available at: www.chumbeisland.com
Goodwin, H.J., Kent, I.J., Parker, K.T. and Walpole, M.J. (1997) *Tourism, Conservation & Sustainable Development*, Vol. 1, *Comparative Report*. Final Report to the Department for International Development (UK).
Halpenny, E. (2002) *Marine Ecotourism: Impacts, Guidelines and Best Practice Case Studies*. The International Ecotourism Society, Burlington, Vermont, 100 pp.
Hawkins, D., Epler Wood, M. and Bittman, S. (1995) *The Ecolodge Sourcebook for Planners and Developers*. The Ecotourism Society, Bennington, Vermont.
Hinrichsen, D. (1998) *Coastal Waters of the World: Trends, Treats and Strategies*. Island Press, Washington, DC.
Hoyt, E. (2000) *Whale Watching 2000: Worldwide Tourism Numbers, Expenditures, and Expanding Socioeconomic Benefits*. International Fund for Animal Welfare, Crowborough, UK.
Kay, R. and Alder, J. (1999) *Coastal Planning and Management*. Routledge, New York.
Kelleher, G. and Kenchington, R. (1991) *Guidelines for Establishing Marine Protected Areas*. IUCN in collaboration with Great Barrier Reef Marine Park Authority, Gland, Switzerland.
Kenchington, R. (1992) Tourism in coastal and marine environments – a recreational perspective. *Ocean and Coastal Management* 19, 1–16.
Lindberg, K. and Halpenny, E. (2001) *Protected Area Visitor Fees – Summary*, 6th edn. The International Ecotourism Society website (www.ecotourism.org).
Mehta, H. and Baez, A.L. (2002) *International Ecolodge Guidelines*, final draft. The International Ecotourism Society, Burlington, Vermont.
Molina, C. and Rubinoff, P. (1998) *Guidelines for Low-impact Coastal Tourism Development – Quintana Roo, Mexico* (Draft Summary – Summer Institute) Amigos de Sian Ka'an A.C., Coastal Resource Center/URI and US Agency for International Development, Rhode Island.
Nelson, V.M. and Mapstone, B.D. (1997) Executive Summary. *A Review of Environmental Impact Monitoring of Pontoon Installations in the Great Barrier Reef Marine Park*. Technical Report No. 13. CRC Reef Research Centre website at: www.reef.crc.org.au/publications/techreport/TechRep13.shtml
NOLS (1998) *Leave No Trace Outdoor Skills and Ethics: Temperate Coastal Zones*. LNT Skills & Ethics Series, Vol. 5, 2nd edn. National Outdoor Leadership School, Lander, Wyoming.
Office of National Tourism (*c*. 1999) *Tourism Facts: Coastal and Marine Tourism*. A pamphlet by Industry Science Resources, Office of National Tourism (Australia), Canberra.
Ryan, C. (1991) The ecological impacts of tourism. In: *Recreation Tourism: a Social Science Perspective*. Routledge, London, pp. 95–130.

Schultz, E.G., McCool, S.F. and Kooistra, D. (1999) *Management Plan Saba Marine Park*. Prepared for the Saba Conservation Foundation, Saba, Netherlands Antilles.

Spruill, V. (2001) National survey of Americans' attitudes toward protected areas in the ocean. *Abstracts: Melding the Science and Policy of Marine Reserves. AAAS Session: Science and the Biosphere, 17 February 2001*. (Sea Web website at www.seaweb.org)

Surfrider Foundation (2001) *Surfrider Foundation*. Homepage of the Surfrider Foundation USA, online. Available at: www.surfrider.org

Sweeting, J., Bruner, A.G. and Rosenfeld, A.B. (1999) *The Green Host Effect: an Integrated Approach to Sustainable Tourism and Resort Development*. Conservation International, Washington, DC.

Tisdell, C. and Broadus, J.M. (1989) Policy issues related to the establishment and management of marine reserves. *Coastal Management* 17, 37–53.

van't Hof, T. (1996) Revenue generation for marine parks. Discussion paper. Available from the author at: tomvanthof@hotmail.com

Viles, H. and Spencer, T. (1995) *Coastal Problems: Geomorphology, Ecology and Society at the Coast*. Edward Arnold/Hodder Headline Plc, London.

Wachenfeld, D.R., Oliver, J.K. and Morrissey, J.I. (1998) *State of the Great Barrier Reef World Heritage Area 1998*. Great Barrier Reef Marine Park Authority, Townsville, Queensland.

Walpole, M.J. and Goodwin, H.J. (2000) Local economic impacts of dragon tourism in Indonesia. *Annals of Tourism Research* 27(3), 559–576.

Wilkinson, C. (ed.) (2000) *Status of the Coral Reefs of the World: 2000*. Australian Institute of Marine Science, Townsville, Queensland.

Witherington, B.E. and Martin, R.E. (1996) *Understanding, Assessing and Resolving Light-Pollution Problems on Sea Turtle Nesting Beaches*. A Florida Marine Research Institute Technical Report No. TR-2. Florida Department of Environmental Protection, St Petersberg, Florida.

WMB Oceanics Australia and Claridge, G. (1997) *Guidelines for Managing Visitation to Seabird Breeding Islands*. Great Barrier Reef Marine Park Authority and Environment Australia – Biodiversity Group, Townsville, Queensland.

CHAPTER 11

The Economics of Tourism in National Parks and Protected Areas

R. Neil Moisey

Introduction

The fact that national parks and protected areas attract visitors that support local economies is undisputed; one must only look at the phenomenal growth that has occurred in many US national park 'gateway' communities such as Gatlinburg, Tennessee, gateway to the Great Smoky Mountains National Park or Jackson, Wyoming, gateway to the Grand Teton National Park. Protected areas from the northern Honduran coast to the South Island in New Zealand also support local economies in communities situated in close proximity to them and to transport links. These communities provide many of the needed goods and services to visitors, and can, if integrated with the management of the natural areas, protect the natural resources for which these areas were set aside. In addition, these local communities can and do provide both economic and political support for the protection and management of these parks and protected areas. Perhaps most important is the role that park visitors might play in this context. In the USA during the early part of the 20th century, rail travel to the national parks introduced the nation to their natural wonders. These visitors became the base of political support for the continued setting aside of additional parklands.

While the role that local communities play in park or protected area management is more fully discussed in Chapter 8, many current issues focus on the economic relationship between parks, protected areas and the communities and businesses that surround them (or are located within, such as Banff or Jasper National Parks, Canada). These communities have as their objectives maintaining the quality of life for their residents, providing economic opportunity, and protecting the values and resources most important to those living there. It is evident that to achieve these objectives, communities with economic ties to parks and protected areas share in the protection and maintenance of these resources (Fig. 11.1).

Clearly, park and protected-area-based tourism is a large and growing part of the economy of many countries. Kenya, Tanzania and Botswana, for example, all have park tourism as their most important export industry. Eagles *et al.* (2000) report that park tourism in the USA and Canada in 1996 constituted an economic impact of between US$236 and 370 billion for that year. The importance of these parks and the related tourism income can be even greater for small local areas near the parks. Typically,

Fig. 11.1. Local businesses near parks may provide specialized travel experiences and programmes to park visitors. Algonquin Outfitters provides wilderness canoeing trips, equipment and training for visitation to Algonquin Provincial Park in Canada. The jobs and employment provide valuable economic benefits to local people. Outfitter's exhibit at the Toronto Sportsmen's Show. (Photographed by Paul F.J. Eagles.)

these are poorer natural resource- or subsistence-based economies that have not kept pace with the economic surge in the later part of the 20th century. Tourism based on these parks might provide the only economic opportunity for those still living in these areas.

This chapter discusses the economic role that parks and protected areas play in regional economies. The economic role that tourism does and will play provides an understanding of its scale in a global sense. Nature-based tourism, or ecotourism, is becoming one of the fastest growth areas and has broad implications for international tourism and for the natural areas that provide the attraction for this tourism sector. The chapter then looks at how visitation to parks is translated into local economic activity and how this is measured. The complexity of local and regional economies can play a significant role in how productive a tourism-based economy might be in addressing economic development goals. The measurement of the economic activity associated with parks provides a conceptual understanding of how communities or countries might change development policies to protect both the parks and the affected communities. Several examples of the significance of park-based tourism will illustrate the varying success of this type of economic development scenario and provide insight into how communities can benefit economically from tourism while sustaining their quality of life.

The Economic Relationship of Tourism, Communities and National Parks and Protected Areas

Nature-based tourism as an economic driver

Tourism is the world's largest growth industry with receipts from international tourism increasing by an average of 9% annually for the past 16 years to reach US$476 billion in 2000 (WTO, 2001). During the same period, international arrivals rose by a yearly average of 4.6% to reach 698 million in 2000. The WTO forecasts that international arrivals will top 1 billion by 2010. Likewise, earnings are predicted to grow to US$1550 billion by 2010.

The linkage between environment as an attraction and economic impact can be substantial. Nature-based tourism, a special niche market, involves travel to unspoiled places to experience and enjoy nature (Honey, 1999). In 1990, nature-based tourism was estimated to account for about 7% of tourism expenditures (Ceballos-Lascuráin, 1993). In 1998, the WTO reported that nature-related forms of travel accounted for approximately 20% of international travel (WTO, 1998). The phenomenal growth of nature-based tourism can be more dramatic for smaller, isolated destinations. For example, in Vietnam, approximately 43% of all tourists visited natural and protected areas in 1998, a 50% increase since 1994 (Koeman *et al.*, 1999). In Montana, USA, it is estimated that about half of the economic impact attributed to tourism was due to recreation activities occurring in wildland settings (Yuan and Moisey, 1992). And in the USA in general, a 1998 study found that almost 50% of vacation trips included nature-based activities; of these, visiting parks or protected areas was the most commonly cited activity (Bruskin Goldring, 1999).

Progression from extraction to service

The economic drivers in natural area proximate communities historically are tied to resource-extractive uses such as timber, mining or agriculture, or subsistence use by locals. These communities tend to exhibit homogeneous economies largely driven by the predominant natural resource sector. These economic sectors have, until more recently, exhibited fairly stable (albeit generally poorer) economic patterns over time. Recent patterns of economic globalization are changing the fortunes of these natural resource-based economies. Much of the value of the resources is processed or added elsewhere resulting in diminishing economic returns to these areas. In addition, large-scale global events, such as economic recessions, produce volatile demand and prices for natural resource commodities (e.g. metals/timber prices). The rapid fluctuation in commodity prices results in unstable socio-economic conditions in these communities.

The introduction of tourism is seen as an economic diversification strategy to reduce reliance on a singular economic sector, to capitalize on the designation of parks and protected areas, and to provide economic incentives for locals to protect the resources in these parks. As the local economy integrates tourism into its economic base, reliance on employment in the traditional economic sectors declines as newer service-based employment provides employment growth. However, workers trained for traditional extractive employment may not possess the skills (nor want to learn them) for employment in the tourism or new service sectors. In-migration by non-locals (and sometimes foreigners) to fill the employment needs of the newly emerging economy can sometimes result in rapid social change within these communities.

Gateways to national parks and protected areas

Park and protected area proximate communities become the 'gateways' to these natural areas (see Fig. 11.2). Gateway communities play an important role in the protection and management of these natural areas in several key ways. First, by providing the needed services for visitors, gateway communities can concentrate the development in the best locations. One example is to locate the developments just outside the park, as is the case with many of the national parks in the USA (e.g. West Yellowstone, Yellowstone National Park; Gatlinburg, Great Smoky Mountains National

Fig. 11.2. Whooping crane, an endangered species, visits the wetlands of the Texas coast each winter. Local tour operators provide viewing opportunities for birders. Such wildlife-related tourism provides valuable employment and income to local citizens. Gateway communities provide a full range of services to the wildlife tourism industry that has developed because of the National Wildlife Refuge. Wildlife tourism at Aransas National Wildlife Refuge, USA. (Photographed by Paul F.J. Eagles.)

Park; and Springdale, Zion National Park), and in other places such as the village of Lukla, just outside Sagarmatha (Everest) National Park in Nepal. This provides the benefits of keeping the parks in a more natural state and allowing the private sector to provide the services and benefit from the tourism-generated economic activity. For example, to tour Zion National Park, visitors are required to leave vehicles outside the park in the community of Springdale and ride shuttle buses into and through the park. Parking and other tourist services are offered in Springdale. One criticism of this approach is the stark contrast that develops at the border of the park and the community. Heavy commercial development can be found within feet of the national park boundaries (e.g. Gatlinburg, Tennessee, borders Great Smoky Mountains National Park, and Estes Park, Colorado, borders Rocky Mountain National Park). Gateway communities might be located many miles away from the park or protected area as is the case with Yulara, Australia, the service city for Uluru National Park–Ayers Rock (located some 20 km away). Yet another model is that of Banff and Jasper National Parks in Alberta, Canada. Each of these communities is located within Banff and Jasper National Parks. The location of gateway communities within the park boundaries can be problematic in terms of ecological and biological impacts and development pressures.

Second, gateway communities can provide economic and political support for the protection of park and protected area resources in several ways. Communities with financial ties to parks (park-based tourism for example) have an inherent interest in the protection of the resource and how the park or protected area is managed. Indeed, the quality of the parks is the primary tourism product. Maintaining the quality of the tourism product is the sustainable approach to tourism development. Visitors can provide both financial and political support for park and protected area management. Gateway communities can engender this by ensuring satisfying experiences for visitors.

Economic Impacts of Tourism in Parks

Types of economic measures

When discussing the economic contribution of tourism in parks and protected areas, two distinct but allied economic concepts can be considered: economic value and economic impact. Economic value measures the broader social benefit derived from designating and protecting parks within a country. During the creation of protected areas, benefits must be weighed against the opportunity costs of other land use options, and thus be reported in a benefit–cost analysis for land allocation decisions. These values can be directly compared with the value derived from other uses of these resources (such as agricultural or extractive use) in making broad policy decisions. Total economic value of the park includes the economic value of use plus non-use values.

The measurement of use values for market-priced goods (i.e. goods and services that are priced through market mechanisms) is relatively straightforward. Estimating prices for non-market goods (such as visitation to parks) is more difficult as no negotiated pricing structure exists. Economists have developed alternative techniques to estimate values for use and non-use values of parks. The two most commonly employed techniques are the 'travel cost method' (TCM) and the 'contingency valuation method' (CVM). The TCM measures use value and is based on visitor expenses incurred in traveling to parks to infer the value of their visit. The CVM can be used to estimate use and non-use values and is based on visitor responses to a set of hypothetical scenarios. These scenarios outline several realistic prices that can be used to infer visitor willingness to pay for a visit to the park or for the protection of the park resources.

The measurement of non-use values is a somewhat more complicated process as users, potential users and non-users hold these values. This involves surveying each of these groups to measure the full range of non-use values. These values include the benefits that accrue to the country of knowing the parks are protected (existence value), that citizens have the option to visit their parks (option value) and that these parks will be protected for future generations (bequest values). These values accrue to both those that visit these parks and those that may never visit. For a more detailed discussion of non-use values and their measurement, see Walsh (1986).

Economic impact focuses more on the local flow of goods and services within the economy. The economic impact generated by parks is primarily due to visitation and spending within the park and surrounding area by tourists and can also occur when a park agency buys supplies within the local area. Economic impact measures changes in sales, employment and income in the region proximate to a park or protected area. It is the generation of income directly derived from tourism that can provide jobs for local communities, profit for local entrepreneurs, tax revenues for local governments and the political and financial support for park protection.

Understanding economic impacts

Tourists visiting parks and protected areas purchase a variety of goods and services from both local and non-local businesses. These expenditures are used to determine 'expenditure profiles', which describe the composition and magnitude of tourist spending. An example of a tourist expenditure profile comes from visitors to Alberta, Canada, national parks, where 39% was spent for lodging costs, 23% towards meals and refreshments, 13% for retail and other expenses, 10% for vehicle rental, 10% for recreation and entertainment, and 5% for local transport (AED, 2000). While an expenditure profile may give an indication of the impact of this direct spending, it does not provide a complete accounting of the total economic impacts produced. The economic effect of each dollar does not stop with the initial expenditure but results in additional ripple effects through the economy.

Figure 11.3 illustrates the flow of money through a regional economy as a result of a tourist purchasing goods or services. These purchases affect not only the businesses directly providing these goods or services, but also those businesses that provide the inputs to produce the goods or services (i.e. labour, raw materials). These purchases then filter throughout the economy, resulting in additional transactions. Each time a transaction is made, income is generated and labour employed in each affected business sector of the economy. If goods or services are not available in a region (a county, a state or a country are examples of an economic region), some transactions may be 'leaked' out to other regions as import purchases. These 'leakages' occur when the payment for goods and services, interest, profits, rents and taxes are outside the region.

The introduction of 'exogenous' money (from outside the region) provides an economic impact in the region. Any exogenous spending in the region will be distributed throughout the economy, producing a multiplier effect on the original expenditure. Since tourists inject outside money, the total effects of this spending on the local economy produce an impact greater than the initial expenditure. Money spent by local

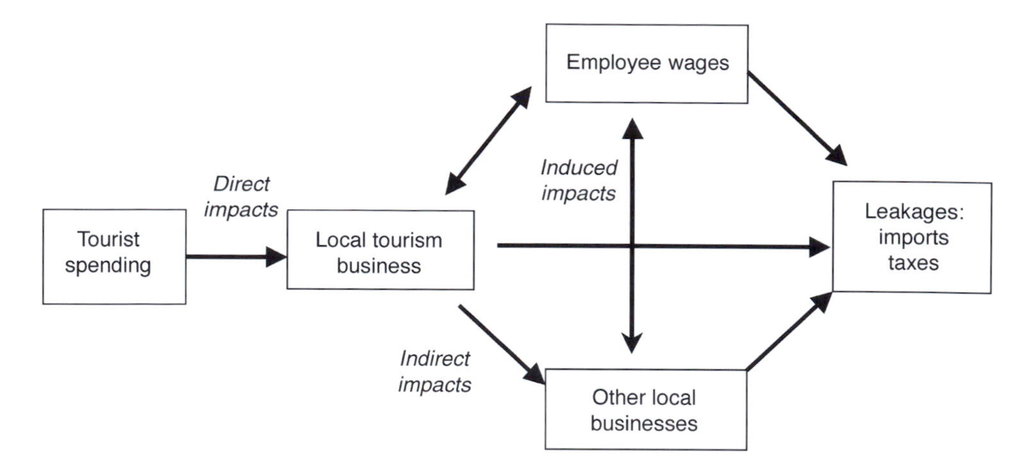

Example: A tourist visiting a park spends $100 in a local hotel. This results in $100 of direct impact in the local economy. The hotel spends money for labour (hotel clerks, gardeners, managers, etc.) who then spend money in the local economy (rent, groceries, etc.), resulting in induced impacts. The hotel also buys locally supplied goods and services (laundry suppliers, business services, building contractors, etc.) that result in indirect impacts. In addition, the hotel buys goods and services (imports) and pays taxes outside the local area, resulting in leakages.

Fig. 11.3. The economic impact of tourism on a local economy.

residents does not produce any economic impact, as this spending is only a redistribution of money already in circulation in the local economy.

There are three types of impacts that 'exogenous' money can have on a region's economy: direct, indirect and induced. Direct impacts result from the initial purchase of goods and services by tourists such as money spent on a local guide service. The local guide purchases inputs (e.g. food and labour) from local suppliers; these purchases result in indirect effects; that is, 'backward linked' suppliers are indirectly affected by the tourist expenditure. Induced impacts result from the increased spending of employees in the directly and indirectly affected businesses. This chain of buying and selling continues until the original expenditure totally 'leaks' out of the region in the form of purchases, interest payments, non-local profits, and rents and taxes paid outside the region.

The indirect and induced impacts are defined as secondary impacts. Total economic impact is defined as the sum of the direct and secondary impacts. The ratio of the direct impact to the direct and secondary impacts is called a multiplier.

$$\text{Multiplier} = \frac{\text{direct impacts} + \text{indirect impacts} + \text{induced impacts}}{\text{direct impacts}}$$

Multipliers give an indication of how much money turns over in the economy and the extent of 'leakage' that occurs from a region as a result of spending. The more that money 'leaks' outside the local economy the smaller the secondary impacts and the smaller the multiplier. Understanding this is especially important for most smaller park proximate communities, as the local economy is generally simple in structure with few if any secondary impacts occurring. The multiplier for a region with a broader diversity of economic sectors will be larger because regional demand may be satisfied from within the region, rather than through imports. Multipliers for large complex regional economies can be as high as 2–3. In the state of Montana, USA, a fairly large area (38 Mha), multipliers for nature-based tourists are around 2.8. Cruise-ship tourism in Jamaica creates multipliers of around 1.2 and around 1.7 in the Dominican Republic. For a smaller, less developed economy with limited linked industries and high rates of leakage, multipliers might be very close to 1.

Multipliers can be calculated for numerous economic indicators. The ratio of direct impact to secondary impact is called an impact multiplier. Just as additional employment earnings are generated as a result of direct expenditures, additional employee compensation is produced from secondary spending. The ratio of direct employee income to direct and secondary employee income is called a personal income multiplier. Employment is generated by each level of impact, producing an employment multiplier, defined as the ratio of direct employment to direct and secondary employment.

Measuring economic impacts

There are many methods and models available to measure the economic impacts of tourism. Primary expenditure data from both park visitors and the spending of the park or natural area agency or organization will be needed. Visitation figures to the park when combined with visitor expenditure data are used to estimate total visitor spending. Therefore, surveys of visitors, local business owners, public agencies and residents are often needed to identify expenditures and business linkages. In larger economic regions, economists have developed economic models that provide estimates of the secondary impacts of tourism spending.

For smaller areas with non-diverse economies, the secondary impacts are likely to be quite low. But, even here, if the area is isolated from other economic regions, the linkages between businesses might be quite strong and secondary impacts might be relevant. In smaller areas there is usually a lack of data available to model the interactions between economic sectors. An additional data concern in smaller, isolated areas is the use of barter systems rather than the transactions based on the exchange of money. Modelling these types of exchange within an economic impact model would require the estimation of the value of goods and services traded.

The economic impact of tourism spending is complex to measure because the economic activity generally does not occur within a single industry but is divided among many. Typically, the economic sectors affected by tourism spending, such as retail or services, also derive considerable income from other non-tourism expenditures. For example, the lodging industry typically accounts for about 30% of tourist spending but also serves unrelated business visitors and non-tourist visits. It is this mixing of inputs (i.e. dollars spent in each sector) that complicates the analysis of tourism's role in an area's economy.

In addition, numerous approaches have been developed to estimate the economic impact that results from tourism spending. These approaches range from relatively simple estimations of direct travel expenditures to sophisticated and costly survey-based input–output (I/O) models or econometric approaches that integrate the most recent advances in the economic modelling field.

Input–output model Output multipliers for a region can be estimated by input–output economic models (see Box 11.1 for an example). Input–output analysis is a method by which the flows of production can be traced through the various sectors of the economy and derives its name from the economic interrelationships within an economy. These models are based

Box 11.1. Measuring economic impact: the Money Generation Model (MGM).

In 1990 the US National Park Service developed an economic model that could be used to estimate economic impacts of parks for local economies. This model was called the Money Generation Model, or MGM. The original MGM focused primarily on the economic impacts associated with park tourism expenditures.

The MGM was subsequently expanded to include the economic effects of two additional types of expenditures, namely expenditures by the Federal Government for National Park Service salaries, park construction projects and other park-related activities; and expenditures by other outside parties, such as state spending for park access roads, or dollars spent by outside interests for marinas, motels, restaurants and other park-related capital development projects.

In 2000, a new version, called MGM2, expanded on the original by providing estimates of the impacts that park visitors have on the local economy in terms of their contribution to sales, income and jobs in the area. The MGM produces quantifiable measures of park economic benefits that can be used for planning, concessions management, budget justifications, policy analysis and marketing.

The updated MGM2 is a spreadsheet-based model that gives the user a wide degree of flexibility in terms of data inputs. For applications with limited visitor spending or local economic data, MGM2 provides sets of default or 'generic' values that can be tailored to a particular application. MGM2 spending data are based on recent NPS visitor surveys so they represent park visitors, rather than general travellers. MGM2 multipliers are based on IMPLAN Pro 2.0 input–output models for local regions around park units.

Continued

Box 11.1. *Continued.*

Model inputs:

- The number and types (segments) of visits/visitors (expressed in person or party nights in the area).
- Average spending for each segment per day or night in the area.
- Multipliers and economic ratios for the region around the park.
- State and local tax rates (optional if tax impacts are desired).

Model outputs:

- Total visitor spending in the local area by visitor segment and spending category.
- Direct effects of this spending in terms of sales, income, value added and jobs in the local area by economic sector.
- Total sales, income, value added and jobs in the region resulting from the visitor spending.
- State and local tax receipts.

The MGM2 and all supporting documentation is available for downloading at: www.nps.gov/planning/mgm/

on the linkages between business sectors within a regional economy. It captures not only the direct effects of a transaction, but also the important and often greater secondary impacts. The method cannot just be used to measure current economic impacts, but through simulation modelling, can also measure impacts of alternative policies.

Input–output analysis involves the development of an input–output table for an economy based on these linkages. The output of sector A becomes the input for sector B whose output then becomes the input for sector C. These relationships can be measured and put into a table format. The framework of the I/O model is the transactions table, producing industries form the rows, consuming industries form the columns of the table. This matrix of interrelationships allows the tracking of goods through the economy.

Economic base model The economic base model aggregates the economy into two sectors, basic and non-basic. The underlying assumption of the economic base model is that a region's economy is driven by its ability to export to the rest of the world. Sales of goods and services from the region to other areas constitute the region's economic base. These industries are called the basic sector. Their sales generate 'new' or exogenous dollars that are injected into the local economy. Employment and income in the basic sectors are a function of this exogenous demand. If external demand for the region's products is reduced (e.g. national recession), then the basic sector output, employment and income are proportionately reduced. Numerous supporting goods and services are required to provide the needs of the basic sectors (e.g. food stores, business services, etc.). These support services constitute the non-basic sector. The size of the non-basic sector is a function of the size of the basic sector. Therefore, any changes in demand for basic sector output will affect the non-basic sector.

The ratio of total to basic employment (or income) forms the employment or income multiplier. This multiplier can be used to model changes in the regional economy and estimate the effects of an influx of 'exogenous' money into the local region. For example, a 10% increase in the output of the basic sector for a region (10% increase in tourism spending) with 100 employed in the basic sector and with a multiplier of 3, will result in the gain of 30 jobs within the region.

Econometric models Econometric models are forecasting methods based on time series data. Econometrics combines statistics, mathematics and economic theory to produce a number of simultaneous equations relating certain economic relationships within the economy. Models can contain up to 200 simultaneous equations to predict changes within a regional economy. These econometric models comprehensively describe the workings of a regional economy, typically forecasting such variables as wages, prices, income and output. The limitations of these models are that very few models have been estimated for small areas owing to the unavailability of data (Glickman, 1977) and the data collection requirements of the econometric model are costly in terms of time.

Examples of the Economic Impacts of Tourism

There are numerous examples of studies estimating the economic impact of tourism on parks and protected areas (Fig. 11.4). The scale of economic impact differs from case to case and can be explained to some degree by the park's ability to attract visitors, the proximity of the park to population centres or transport linkages and the ability of the local communities to service the needs of tourists. Each of these factors can play a significant role in the success of sustaining both a local tourism-based economy and the park resources themselves. The following cases provide a broad overview of the economic impact generated by parks in a variety of settings across the globe.

Australia

In a study of the economic impact of tourism in five Australian World Heritage Areas (Great Barrier Reef, Wet Tropics, Uluru National Park, Kakadu National Park and Tasmanian Wilderness), it was estimated that tourism expenditures in 1991–1992 were Aus$1,372,000,000. The total management budgets were Aus$48,700,000, and the user-fee income to the management agencies was Aus$4,160,000 (Driml and Common, 1995). These protected areas generated economic impacts that were well

Fig. 11.4. Research into the expenditures of the birders at Point Pelee National Park alerted the local community of Leamington to the economic significance of this activity. The business community became much more interested and supportive of park management on learning that birders spent millions of dollars locally each year. Birders in Point Pelee National Park, Canada. (Photographed by Paul F.J. Eagles.)

over 28 times the costs of managing these protected areas. At the same time, user-fees only generated about 8.5% of the total management costs for these areas.

In another study conducted for the National Park and Wildlife Service, Christiansen and Conner (1999) estimated the economic impact of the contribution of the nature reserve to the regional economy of the Eurobodalla Local Government Area. An input–output analysis was conducted based on the operating budget of the nature reserve and the tourist expenditures within the region. The NPWS management expenditure resulted in Aus$233,000 in gross regional output. This represented a multiplier of 1.92, indicating that for every dollar spent by the NPWS on park management, another Aus$0.92 in gross regional output was generated elsewhere in the local economy. Surveys found that expenditures by visitors to the island contributed an estimated Aus$1,400,000 in gross regional output per year to the regional economy. This included Aus$468,000 in household income paid to 19 people in the local economy. The aggregated NPWS and visitor expenditure impacts were estimated to be Aus$1.65 million in gross regional output and Aus$857,000 in gross regional product, including Aus$588,000 in household incomes, which equates to 26 local jobs. It is important to note that this impact occurred with quite modest numbers of park visitors, indicative of the fact that even small numbers of visitors can have important local economic impact. The communication of such important economic impacts can influence local citizens' attitudes towards the park and its management. When people understand the extent of the economic contribution of these parks they are more willing to support the continuance of these programmes.

Case Study Number 7: Montague Island Nature Reserve (Australia)

Tourism Planning, Impact Monitoring and Community Development

Montague Island is off the south-eastern coast of Australia. The reserve is a unique example of a protected area agency using tourism to provide essential financial support for conservation and the local community in an environmentally and politically sensitive protected area.

The nature reserve contains both natural ecosystems (penguins, seals, sea birds) and cultural features (European and Aboriginal history) of national importance. The island became a Nature Reserve in 1990 and was placed under the care and control of the New South Wales National Parks and Wildlife Service.

From 1990 to the present, the New South Wales National Parks and Wildlife Service developed a system of careful carrying capacity determination, community consultation and monitoring of impacts. Measurement of the economic impact of the tourism showed the value of financial impact monitoring.

The park agency encourages research by university faculty and students. Research is ongoing on a wide variety of topics, ranging from shearwater biology to solar energy utilization. Research shows increasing numbers of Australian fur seals on the island, for example. The number of visitors is carefully controlled to ensure suitable levels of economic and environmental impact. Much of the island is off-limits to visitors, with access limited to well-marked trails in accessible areas.

In 1999, the Montague Island Nature Reserve was given the British Airways/World Conservation Union Award for Park Tourism. This award was given because of the following features of this project.

1. Careful development of a capacity limit for use.
2. Thorough financial management so that tourism use pays for the essential management elements of the park.
3. Good cooperative arrangements with the local community and with local universities.
4. An openness to continual monitoring and evaluation of all aspects of park operation.
5. A highly professional economic impact study.
6. Good integration of both cultural and natural heritage elements in the tourism programme.
7. National ecotourism accreditation at the highest level.
8. An obvious professionalism by the government agency staff members in all aspects of tourism management.
9. The use of tourism to protect and manage some very important biological features.
10. The ability of the site to serve as a best practice example for park tourism managers elsewhere.

Website: www.npws.nsw.gov.au/parks/south/sou018.html

Belize

Lindberg and Enriquez (1994) report on the economic impact of ecotourism in Belize. Tourism is a major component of the Belizean economy. The study estimated that tourism generated US$211,000,000 in total impact and US$41,000,000 in personal income. One significant finding of the authors' report was that positive conservation attitudes and protected area support increased as ecotourism benefits accrued to communities.

Canada

In a 1998 study of the economic impact of visitors to the Rocky Mountain National Parks in Alberta (Banff, Jasper and Waterton), it was estimated that tourists spent over Can$954 million in the province that resulted in Can$640 million of secondary impacts and generated 28,000 jobs (AED, 2000). These parks have international prominence and their geographic concentration adds to their attractiveness as a nature-based tourism destination. In addition, they are easily accessible from the Asian market.

USA

The National Parks and Conservation Association estimated that visitors to Utah's national parks generated more than US$554 million in tourism sales and just over 30,000 jobs in the state's economy in 1995 (Voorhees *et al.*, 1996). The nationally famous parks are mainly located in the remote southern section of the state and attract visitors from across the USA and internationally. The study estimated the impacts at the state level, which included the impacts that occurred in the major metropolitan area, Salt Lake City.

Tourists visiting Grand Teton National Park in Wyoming provided about 27% of the income and employment in rural Teton county (Merrifield and Gerking, 1982). The mid-sized community of Jackson Hole, Wyoming, is located within a few miles of the border of the park and is serviced by a major airport within minutes of both the park and the community.

Visits to Grand Canyon National Park, USA, result in a sizeable economic impact in the local rural area: US$30 million of total economic impact in the community of Williams, Arizona (Leones and Ralph, 1997). But studies have shown that the majority of visitor expenditures are made in the larger urban centres within a day's drive of the park.

Yuan and Moisey (1992) report that in Montana approximately 44% of the tourism industry receipts were attributed to tourists recreating in wildland settings. Tourism is the second largest industry in the state. Almost half of Montana tourists visited either Glacier or Yellowstone National Park while in Montana. Each of these national parks has international prominence and they are major tourist attractions for the state. While most of the economic impacts generated by tourists visiting these parks are concentrated in the park proximate communities, the economic impacts are distributed throughout the state as tourists journey through Montana to and from their destinations.

Discussion

National parks and protected areas are powerful generators of economic impact. Nature-based tourism can provide both economic opportunity and protection of park resources. Park proximate communities and parks can benefit from the mutual relationship based on the sustainable use and management of the park resources. Figure 11.5 illustrates the roles of the three major participants. Park management agencies not only provide the nature tourism opportunities, but also manage and protect the main tourism product. The tourism industry facilitates many of the tourist opportunities and provides an array of supporting lodging, eating and transport services. The local residents, who may benefit from tourism development, may also pay certain costs associated with impacts on quality of life, physical infrastructure and services. Thus, the sustainability of

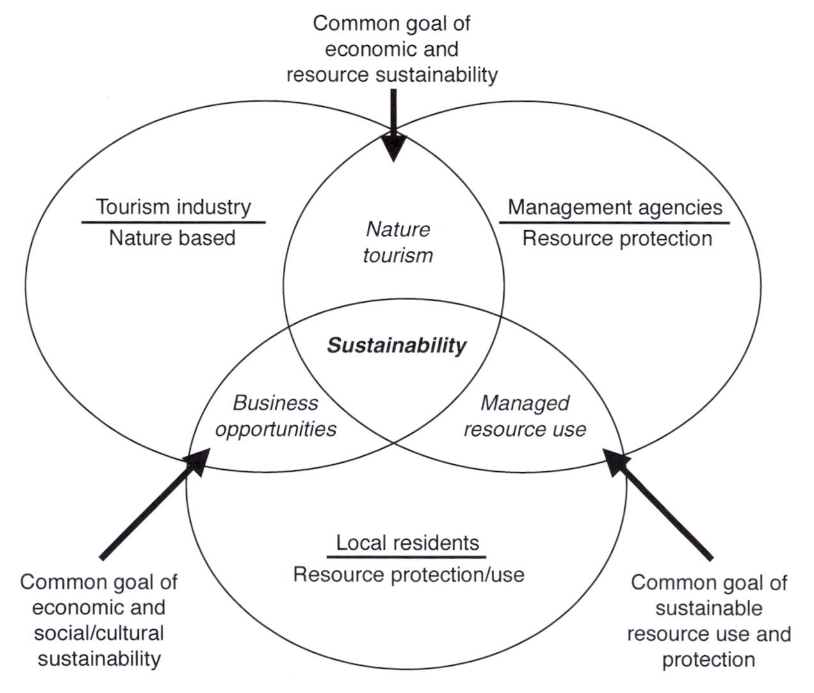

Fig. 11.5. Shared goals for the sustainability of the tourism industry, the community and the park resources.

the tourism industry, the community and the park resources is shared equally within this context. An additional issue is the potential for the disproportionate sharing of costs and benefits of tourism. In many cases, the economic benefits accrue to very few members of the local community while the costs are shared by a larger cross-section.

A common tourism development approach to enhance economic opportunity within the community while protecting the park resource is to increase economic benefits. The traditional demand-driven approach is to attract more visitors. But, given the potential negative impacts (environmental, social and economic) of ever-increasing visitor numbers, other alternatives may be just as appropriate or even more effective.

There are many ways in which to increase the local economic benefits of tourism without having to attract more visitors. A strategy that focuses on the flow of tourist money through the local economy (Fig. 11.3 on page 242) would indicate three areas in which to concentrate efforts. First, increasing the spending per visitor would inject more money into the local economy. This can be accomplished through an increase in available spending opportunities. This could involve the sale of handicrafts, providing guide services, or local accommodation and food preparation.

A second strategy would be to reduce the leakages from the economy. Leakage occurs through the use of imported labour and products, repatriation of profits by non-resident owners, and marketing, transport and other services based in the originating country. Research has shown that the leakages in destination countries can be quite substantial. On average, 55% of tourist expenditures are spent outside the destination countries (up to 75% in Gambia and Commonwealth Caribbean), but this can be as little as 25% for large economies such as India (Ashley *et al.*, 2000). This is even more dramatic in relatively undeveloped economies in remote regions of the world that do not provide much of the needed economic inputs to satisfy local demands. For example, in the Annapurna trekking region of central Nepal it is estimated that only about 7% of the US$3 spent by an average trekker per day is retained within the local economy (PANOS, 1995).

A third strategy to reduce leakages would involve increasing the linkages within the local economy by encouraging local participation in the tourism industry and support services (Fig. 11.6). In Zambia, for example, privatization of game lodges and hunting concessions is regulated through an international competitive bid process, but local investors and entrepreneurs are encouraged through the reservation of certain leases and hunting concessions for local involvement. Increasing local participation in the tourism industry spreads the economic benefits to a larger cross-section of the community resulting in greater retention of tourist dollars within the local area. To accomplish this might involve providing additional financial and training support to local residents.

Fig. 11.6. The public and private reserves provide up to nine levels of pricing of accommodation in and around Kruger National Park. These range from inexpensive campsites, through moderately priced cottages to luxury ecoresorts. This range provides for maximum service level opportunity and maximum income capture. This photo shows medium-priced cabins owned and managed by South African National Parks. Rondovels at Skukuza in Kruger National Park, South Africa. (Photographed by Paul F.J. Eagles.)

References

AED (2000) *The Economic Impact of Visitors to Alberta's Rocky Mountain National Parks in 1998.* Alberta Economic Development, Edmonton, Alberta.

Ashley, C., Boyd, C. and Goodwin, H. (2000) *Pro-poor Tourism: Putting Poverty at the Heart of the Tourism Agenda*, no. 51. Overseas Development Institute, London.

Bruskin Goldring Research (1999) *Nature-based Activities and the Florida Tourist.* Edison, New Jersey.

Ceballos-Lascuráin, H. (1993) Ecotourism as a world-wide phenomenon. In: Lindberg, K. and Hawkins, D.E. (eds) *Ecotourism: a Guide for Planners and Managers.* The Ecotourism Society, North Bennington, Vermont, pp. 12–14.

Christiansen, G. and Conner, N. (1999) *The Contribution of Montague Island Nature Reserve to Regional Economic Development.* New South Wales National Parks and Wildlife Service, Hurstville, New South Wales.

Driml, S. and Common, M. (1995) Economic and financial benefits of tourism in major protected areas. *Australian Journal of Environmental Management* V(2:1), 19–39.

Eagles, P., McLean, D. and Stabler, M.J. (2000) Estimating the tourism volume and value in parks and protected areas in Canada and the USA. *George Wright Forum* 17(3), 62–76.

Glickman, N.J. (1977) *Econometric Analysis of Regional Systems.* Academic Press, New York.

Honey, M. (1999) *Ecotourism and Sustainable Development: Who Owns Paradise?* Island Press, Washington, DC.

Koeman, A., Nguyen Vrn, and Ltr Vrn Lanh (1999) The economics of protected areas and the role of ecotourism in their management: the case of Vietnam. Paper presented at the *Second Regional*

Forum for Southeast Asia of the IUCN World Commission for Protected Areas, Pakse, Lao PDR, December 1999. IUCN Lao People's Democratic Republic Country Office, Vientiane, Lao People's Democratic Republic.

Leones, J. and Ralph, V. (1997) *The Williams 1995–96 Visitor Study*. Department of Agricultural and Resource Economics, University of Arizona Cooperative Extension, Tucson, Arizona.

Lindberg, K. and Enriquez, J. (1994) *An Analysis of Ecotourism's Economic Contribution to Conservation and Development in Belize*. World Wildlife Fund, Washington, DC.

Merrifield, J. and Gerking, S. (1982) *Analysis of the Long-term Impacts and Benefits of Grand Teton National Park on the Economy of Teton County, Wyoming*. Institute for Policy Research, University of Wyoming, Laramie, Wyoming.

PANOS (1995) ECOTOURISM: Paradise gained, or paradise lost? Panos Media Briefing No. 14. Panos Institute, London (www.oneworld.org/panos/briefing/ecotour.htm#Ch2).

Voorhees, P., Mednick, A.C. and McQueen, M. (1996) *The Economic Importance of National Parks: Effects of the 1995–1996 Government Shutdowns on Selected Park-dependent Businesses and Communities*. A Report of the National Parks and Conservation Association. NPCA, Washington, DC.

Walsh, R.G. (1986) *Recreation Economic Decisions: Comparing Benefits and Costs*. Venture, State College, Pennsylvania.

WTO (1998) Ecotourism, Now one-fifth of Market. WTO Newsletter, January/February. World Tourism Organization, Madrid. Available at: www.world-tourism.org/omt/newlett/janfeb98/ecotour.htm

WTO (2001) Tourism highlights 2001 updated. World Tourism Organization, Madrid. Available at: www. world-tourism.org/market_research/data/pdf/highlightsupdatedengl.pdf

Yuan, M.S. and Moisey, N. (1992) The characteristics and economic significance of visitors attracted to Montana wildlands. *Western Wildlands* 18(3), 20–24.

Further Reading

Beeton, S. (1998) *Ecotourism: a Practical Guide for Local Communities*. Landlinks Press, Collingwood, Australia.

Bosselman, F.P., Peterson, C.A. and McCarthy, C. (1999) *Managing Tourism Growth: Issues and Applications*. Island Press, Washington, DC.

Linberg, K., Wood, M.E. and Engeldrum, D. (eds) (1998) *Ecotourism: a Guide for Planners and Managers*. The Ecotourism Society, North Bennington, Vermont.

Machlis, G.E. and Field, D.R. (eds) (2000) *National Parks and Rural Development*. Island Press, Washington, DC.

Power, T.M. (1988) *The Economic Pursuit of Quality*. ME Sharpe, Inc., New York.

CHAPTER 12
Park Tourism: Marketing and Finance

The Role of Finance in Park Management

All parks require money to operate. Facilities, programmes and people must have adequate funding. This fact is self-evident, but strangely is an underdeveloped aspect of park management. This chapter covers park marketing and finance and their critical relationship to tourism.

Typically, park budgets contain two important categories of expenditures: operational and capital. Operational expenditures support the recurring and day-to-day aspects of parks, such as staff salaries, purchase of supplies, payment of utility bills and facility maintenance. Capital expenditures support the enduring goods such as the construction of roads and buildings, the purchase of vehicles and the purchase of maintenance equipment.

Parks obtain their funding from several sources. Government tax-based grants are the most common source of funds. Increasingly, the fees and charges applied to tourists and other users provide a major proportion of funds. Occasionally, donations are important. These three sources, government grants, tourism charges and donations, provide virtually all the income of most parks.

The conservation of natural and cultural resources is generally regarded as a public good, with all of society benefiting. However, the provision of public use is a private good, with only those undertaking recreation benefiting. Therefore, charges for visitor use of a park reflect a balance between the public good and the private good. The relative proportion of the management cost ascribed to the public good, and therefore to taxes, or to the private good, and therefore to user fees, is an important public policy. There is considerable debate concerning the relative proportions of these budget elements (Fig. 12.1).

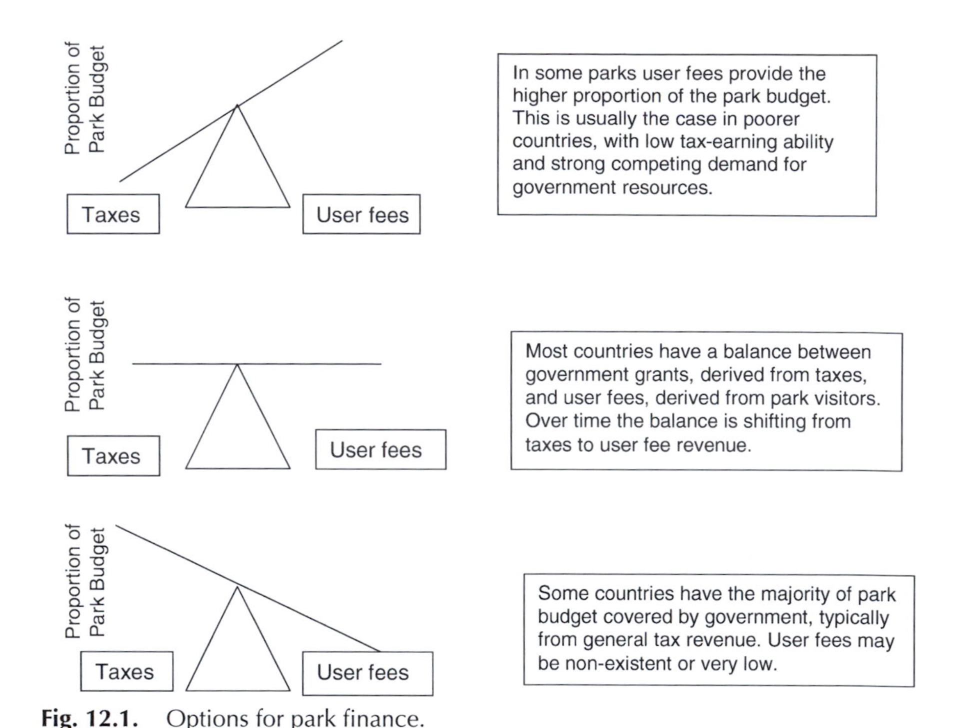

Fig. 12.1. Options for park finance.

The desired outcomes of a user-pays fee income system are assumed to be: (i) cost-effectiveness; (ii) improved park management; (iii) better visitor facilities and services; and (iv) more positive public attitudes towards a park and a park agency. A study in Australia found that all national park agencies reported that the user-pays fee income systems achieved these four outcomes, to various degrees (Queensland Department of the Environment, 2000). The study also reported that the perceived benefits of the user-pays system in Australia included:

- dramatic and visible improvements in park facilities and management;
- increased staffing through user-pays funded positions;
- the establishment of contact between park staff and users, including commercial operators, park visitors and leaseholders;
- a more equitable situation where clients, especially interstate and international visitors and commercial users, pay for services received, rather than the entire burden being carried by local taxpayers;
- greater appreciation by users of services and facilities they pay for; and
- stimulus for employment and small business operators, especially in rural communities.

There are objections to parks charging fees. Many people have a very low reference price for park use, typically based on past experience of low fee levels. They therefore expect their use to be ‘free’. Some people expect others to subsidize the cost of providing the services and facilities that they use. Many people feel that nature is free and ubiquitous and that it is improper to charge money to experience nature. There are concerns that use fees may discriminate against low-income users, or reduce use levels and harm local tourism. Many commercial operators undertake business planning 18 months in advance. They require substantial prior notice of any change or increase in use fees. There are strong concerns among many environmental groups that user-pay systems will lead to commercialization of parks, to the detriment of environmental quality.

In developing countries, foreign aid can be a very important source of funds. For example, in the eastern African countries of Kenya and Tanzania almost all operating funds for the parks and game reserves come from tourism fees and charges. The majority of the capital funds come from foreign aid, often from the European Union and the United States. Major non-governmental organizations (NGOs) can be important as well. For example, the Frankfurt Zoological Society has a long-standing programme of providing operating and capital funding to the national parks of Tanzania.

Public- and private-sector finance is a broad topic with an extensive literature. The World Commission on Protected Areas produced guidelines on park finance for protected area managers (Thomas *et al.*, 2000). That document is a primer on the concept of park finance.

Management Effectiveness

Management effectiveness includes three main components: (i) the goals and design of protected areas; (ii) the appropriateness of management systems and processes; and (iii) the delivery of the services and programmes to fulfil objectives (Hocking, 2000). There is considerable concern that many parks and protected areas are managed ineffectively. One of the most important input factors determining management effectiveness is the adequacy of the available resources, such as park finance. There are other aspects influencing management, such as the staffing levels the staff training levels, and the level of external threat. However, most management problems can be solved if adequate finance is available. Therefore, management effectiveness is heavily influenced by park finance. Successful management of park tourism is only possible with sufficient levels of financial and human resources.

Public and Private Sector Financial Relationships

The public sector has the unique role, based on a societal mandate, of natural and cultural resource protection of park resources (Table 12.1). An important aspect of protection is the determination of acceptable

Table 12.1. Public sector role in park tourism.

Environmental protection
Infrastructure (roads, airports, rail lines, electricity, sanitation)
Security and enforcement
Monitoring of impacts, evaluation of quality
Allocation of access
Limits of acceptable change
Information (interpretation, visitor centres)
Conflict resolution
Guides and recreational programmes
Accommodation (campgrounds, cabins)

uses and use levels. Security of the environment and the public is a government responsibility. Typically, basic tourism infrastructure is paid by the public purse. All the necessary roads, airports, rail lines, electrical distribution networks and sanitation infrastructure are paid for by tax revenue. Security and policing is a government responsibility. The monitoring of impacts on the park and evaluation of the value of these impacts is a government responsibility. When there is conflict over recreational or resource use, government develops procedures for resolution.

The provision of information comes through a combination of the public and private sectors. Many individuals and institutions produce the books, brochures, films and websites that provide information to the public. The public agencies have a responsibility to ensure that accurate and up-to-date information on the park is available. However, this is a difficult and, in some cases impossible, situation when the public agency is small and poorly funded, and the private sector is large and well funded.

In some parks, guides and recreational programmes are provided by the park agency. The staffing of visitor centres, the operation of campfire programmes and the offer of conducted hikes are all examples of programmes that can be supplied by the agency.

Many parks provide campgrounds; a few provide cabins and lodges. Typically, these are operated by government agencies. In this case all the income flows through the government's coffers. Often, these facilities are owned by the park, but are leased to private companies for operation. In this case the majority of the income goes to the private operator, but a portion goes to the facilities' public owners based on contractual and lease stipulations.

In some countries, information is provided by a coordinated effort of both the public and the private sectors. New Zealand is a world leader in the development of a sophisticated, community-based visitor information system for park tourism. This is especially evident on the South Island where most towns and national parks have visitor centres to serve travellers. These are clearing houses for all types of visitor information, including sophisticated levels of accurate information on the services, programmes and facilities in the national parks. The public and private cooperation in New Zealand produces an advanced and appreciated information source for travellers.

In most countries, the majority of the cash flow surrounding park tourism is through the private sector. Typically, the private sector provides most of the services and consumer products used by park visitors (Table 12.2). Private operators provide most of the accommodation, food, transport, media and advertising. The private sector has the ability to respond quickly to consumer demands and to develop specialized products. In poorer countries, such as in most countries in Africa, information is largely provided by private operators.

The private sector relies heavily on the public sector for resource protection, infrastructure and security services. The public sector relies heavily on the private sector for handling the day-to-day activities of the visitors to the parks. A typical example is that of transport. The public sector builds and maintains the transport infrastructure, the roads, rail lines and airports. However, the private sector provides all the buses, tour vans and planes that operate within this infrastructure. There is constant conflict over the payment for these facilities. Many private operators try to function like a parasite, that is, they use the facilities without providing any funds to government for construction and maintenance. It is to the short-term benefit of any one operator to avoid paying for any infrastructure while relying on these facilities paid for by taxpayers at large or by the fees from other operators. Therefore, it is a critical role of the government park agency to develop an equitable and fair system of charges. It is then important to ensure that this system is totally implemented, making certain that all operators pay their appropriate allocation.

Public and private competition is evident in the provision of information on the Internet. Information is widely available in this fashion, including that typically available in visitor centres, in park publications and in guidebooks. Better-funded park agencies have sophisticated websites that provide all the basic information required by an individual park visitor. Poorly funded park agencies either do not have websites, or have

Table 12.2. Private sector role in park tourism.

Accommodation (lodges, hotels, campgrounds)
Food (restaurants and food stores)
Transportation (buses, automobiles, airplanes)
Information (guides, interpretive programmes)
Media (films, books, videos)
Promotion and advertising
Consumer products (clothes, souvenirs, equipment)
Personal services (entertainment)

weak sites containing out-of-date material. There is a plethora of private sites providing park and park tourism information. Private tourism operators often provide web information in considerable depth. NGOs, such as environmental and recreational groups, have sites that explain their point of view. There have even been cases where interest groups have developed Internet websites that are designed to mimic the park agency sites. They provide a modified message that is at variance with agency policy but in line with the policy that the interest group would like to see in the park. Researchers often provide websites explaining their research programmes in parks and the results. This rich source of information is valuable for the potential park visitor.

However, it is important to note some problems with this emerging technology. Park agencies often find it difficult to provide the level of accurate and current information now demanded by computer-literate park tourists. They also find it difficult to monitor the data provided by the private sector, to ensure that it is accurate and not misleading. Hockings (1994) studied the relative roles of the public and private sectors in the training and provision of interpretation in the Great Barrier Reef Marine Park. He found that most tour operators provided information as part of the trip experience, but the vast majority of tour operators did not use professionally trained interpreters. Only one-third used the staff training services provided by the Marine Park Authority. Langholz (1996) found that private nature reserves in Africa and Latin America reported that lack of cooperation with government was a major problem. Clearly, ongoing discussions and monitoring of the interrelationships between the private and public sectors in park tourism is necessary.

Monkey Mia Dolphin Resort in the Shark Bay World Heritage Area in Australia is a leader in public and private tourism cooperation. The Monkey Mia Dolphin Resort is particularly impressive for the assistance given to dolphin researchers, the good relationship with the local resource management agency, and its funding contributions to Project Eden. Project Eden is a major programme involving the restoration of ecosystem integrity in the Peron Peninsula of Western Australia, including the elimination of feral species and the reintroduction of five endangered species. The cooperative relationship between the private and public sectors is a worthwhile goal of ecotourism development but unfortunately does not always develop to its fullest potential. Langholz (1996) reports that Costa Rica is planning to offer private reserves special status in the form of conservation easements and official recognition. This is a gesture to recognize the growing conservation and tourist roles of private reserves in that country.

The operation of an ecotourism industry requires the cooperation of both the public and private sector. Neither can do the job alone. Each is fundamentally dependent on the other. This situation is not always appreciated. Much time and effort is wasted in conflict situations where none fundamentally exists. The long-term health of the natural environment and the financial condition of all sectors of ecotourism depend on cooperation.

The Finance and Marketing of Tourism Management in Parks

What are tourists seeking in their travel? Useful answers come from Canadian studies. Tourism Canada (Burak Jacobson, 1985) found the travel motives of the average Canadian traveller to be socially oriented (Table 12.3). Friends, family, entertainment, safety, predictability and fun are important. Warm, predictable weather is highly ranked, as are locales near lakes and streams. The social orientations are more important than the environmental features of destinations.

In comparison, research by Eagles (1992) showed that the travel motives of the Canadian ecotourists are attraction-oriented (Table 12.4), with tropical forests, wilderness and wildlife highly ranked. Ecotourists are most interested in experiencing, learning and photographing wild nature within natural settings. Filion *et al.* (1993) found that 18.7% of

Table 12.3. Travel motives of Canadian tourists.

Travel motive	Rank
Be together as a family	1
Feel at home away from home	2
Visit friends and relatives	3
Warm climate	4
Have fun and be entertained	4
See maximum in time available	6
Lakes and streams	7
Meet people with similar interests	8
Go to places where one feels safe	9
Predictable weather	10

Table 12.4. Travel motives of Canadian ecotourists.

Travel motive	Rank
Tropical forests	1
Wilderness and undisturbed nature	2
Learn about nature	3
Birds	4
Lakes and streams	5
Trees and wildflowers	6
Photography of landscape and wildlife	7
Mammals	8
National and provincial parks	9
Be physically active	10

Canadians took a trip in 1991 to view or photograph wildlife. These people devoted 84,300,000 days to this activity and spent Can$2.4 billion during the trips. This attitude base and activity profile represents the underpinnings of the ecotourism industry in Canada. The travel motives of other North Americans, Australians and Europeans may be similar to the Canadians, but further research is needed.

Ecotourists' motives are fundamentally different from the typical traveller's motives. As a result, the ecotourism travel industry must be designed differently from the standard travel approach. Research suggests that the key concepts underlying ecotourist travel motivations are wilderness, wildlife, parks, learning, nature and physical activity. In order to satisfy ecotourists, these ideas should underlie the management of natural resources and the provision of tourist services in parks.

The travel experience has five discrete periods (Fig. 12.2). Park-based tourism is unusual in that so much of the travel experience is beyond the scope of the park agency. Much of the recognition of the product, all of the travel to and from the site, and most of the post-trip recollections are influenced by sectors of the economy out of the control of the park agency. Only during the in-park experience does the management agency have some control. This circumstance has considerable impact on the trip experience satisfaction of the tourist. For example, the site manager has little control over the development of appropriate anticipation within the consumers. Since satisfaction with a product is the result of the difference between anticipated benefits and actual benefits, visit satisfaction can be impacted by anticipation that is inaccurate. Park agencies have a major challenge in developing their tourism marketing because of the small portion of the travel experience continuum under their control.

Parks and recreation marketing is substantially different from the marketing of manufactured goods (Table 12.5). Understanding the unique nature of the outdoor recreation product is essential for planning and management. With recreation products, the consumer is actively involved in the creation of the product. This fact requires sophisticated personal service features within the recreation product. The ephemeral nature of recreation requires the creation of memory reinforcements, such as souvenirs, to help the consumer remember and communicate their travel experience. The difficulty in testing the product before purchase makes the travel consumer look widely for quality indicators and accurate information. Recreational products cannot be stockpiled for periods of high demand and cannot be returned if defective, leading to unique problems in supply. These, and other factors, require sophisticated business

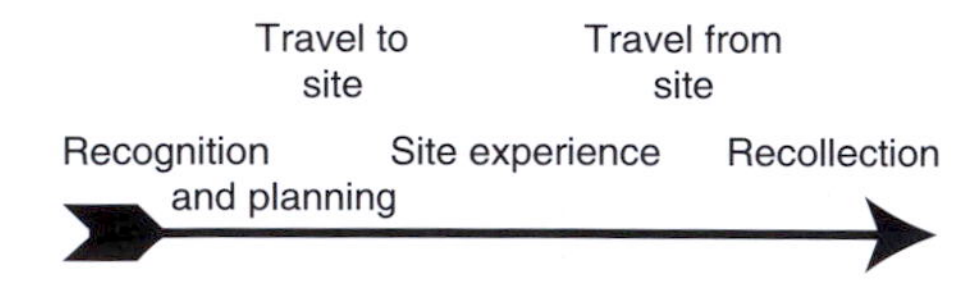

Fig. 12.2. Travel experience continuum.

Table 12.5. Parks and recreation marketing.

Unique aspects of park experience marketing
Outdoor recreation experiences are consumed on a site well away from home
Travel costs to the site often far exceed costs at the site
It is a package of facilities and programmes that attracts people to a site or area
Recreation experiences are ephemeral and experiential and cannot be possessed, except as memories
The production, delivery and consumption of the recreation product occur simultaneously
Consumers are actively involved in the production of their experience, both their own and that of others
Poor recreational experiences cannot be returned for a refund
Recreational sites and experiences are very difficult to evaluate before purchase, therefore word-of-mouth from friends and family is a very important choice determinant
Recreational products cannot be stockpiled during periods of low demand and sold during times of excessive demand
Important aspects of the recreation experience occur before and after the on-site participation

analysis and managerial training if the park tourism business is to be successful. It is important that park tourism employees have specialized training in service marketing and not just in product marketing.

Many park agencies are reluctant participants in the tourism business. Most managers are trained in resource management. Few have professional education in tourism, finance or marketing. As a result, the development of tourism policy is often reactive, with a weak conceptual and policy basis.

The complexity of tourism management in parks is frequently underestimated. Managers must balance environmental protection and visitor use of the resource. Park managers must deal with the demands of visitors, local residents, regional interests, the national government and the private tourism industry. Usually the park is managed as a public good, owned by government and financed from tax revenues with all tourism products sold at an operating loss.

It is critical to recognize that tourism management in protected areas is complex. It requires a sophisticated management structure with well-trained staff.

Park Fees and Pricing Policy

Table 12.6 summarizes the full range of income generation opportunities in park tourism now being utilized by park agencies and their private sector partners in various locales. Most park agencies obtain the majority of their tourism income from entrance fees and charges for facilities, such as campground use. A few have special parking charges, such as for the second car at a campsite. A very few operate their own food and souvenir stores.

Gate fees alone are usually not sufficient to provide all operational income for a park. Even for heavily visited sites, park gate fee revenue rarely covers total operational costs and never covers capital costs. Large gate fees can reduce visitor diversity due to the loss of some low-income groups (see Fig. 12.3). Agency guidelines typically specify standard system-wide fee policies. Given the diversity occurring in parks, fees and charges should be set on a site-specific basis. Therefore, flexibility for any one park within the agency-wide fee policy is necessary. Fee collection

Table 12.6. Park tourism income sources.

Park entrance fees
Recreation service fees, special events and special services
Accommodation
Equipment rental
Food sales (restaurant and store)
Parking
Merchandise sales (equipment, clothing, souvenirs)
Licensing of intellectual property
Donations
Cross-product marketing

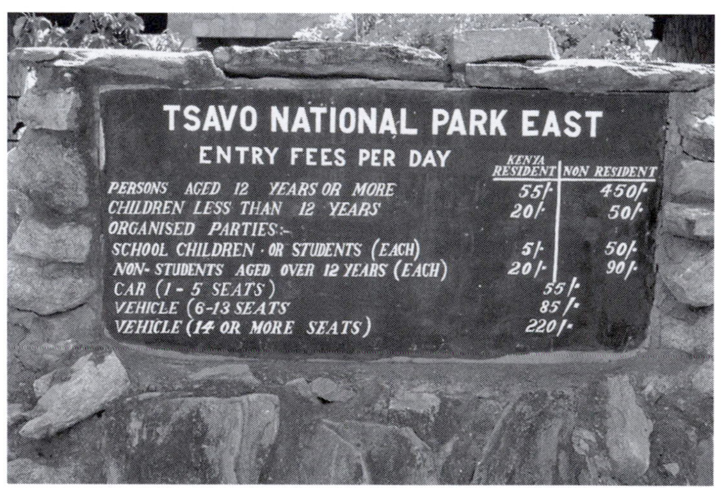

Fig. 12.3. Kenya charges much higher entrance fees for foreigners than for national citizens. This fee structure is designed to earn income from foreign visitors and to encourage visitation by local citizens. However, gate fees alone are usually not sufficient to provide all required operational income for a park. Entrance fees board in Tsavo National Park, Kenya. (Photographed by Paul F.J. Eagles.)

may not be possible at all sites. Those with very low use levels may not justify the cost of fee collection (Lindberg and Huber, 1993). Some parks are experimenting with voluntary entrance payments, typically with a warning that non-compliance may result in a fine. In these situations, park police check for non-compliance at intervals. Surprisingly, only a few parks have active donation programmes. Most people are loath to donate to government but are much more amenable to donate to NGOs. Accordingly, Friends Groups associated with the parks undertake most of the donation programmes.

The fiscal success of one park agency in Canada is worthy of discussion. The Niagara Parks Commission in Ontario operates the public parkland and recreation facilities along the Niagara Gorge and Niagara Falls area of Canada. This agency manages linear parkland of 40 km in length, running from Fort Erie on Lake Erie, north to Niagara-on-the-Lake on Lake Ontario. At the centre of this strip is the magnificent view of the two mighty Niagara Falls. The parkland is composed of formal gardens, significant historical sites, special recreational facilities and small patches of natural woodland and canyon. The Niagara Parks Commission is a special purpose park agency operating under its own legislative authority. A board composed of appointees from both the local municipalities and the province governs it. The agency has been in operation since 1885. This commission has the legislative powers to function like a private municipality or a corporation: it has exclusive authority to provide all the services and programmes within its legislated parkland; it can tap capital markets for loans; it can construct facilities; it can provide whatever services and programmes it sees as necessary; it can set its own prices and can keep all the income. The Niagara Parks Commission is one of the few park agencies in the world that runs an operational profit each year. This is even more remarkable when one sees that it does not charge for it primary tourist attraction, the viewing of Niagara Falls. There are no charges for entering Niagara Parks Commission parkland or for visiting and viewing Niagara Falls.

The financial success of the Niagara Parks Commission lies in the multitude of income sources it obtains from its high-volume tourism. This government agency provides virtually all of the special recreational programmes in the Niagara Parkland. It operates a cable car, a tunnel walk under the Canadian Falls, a butterfly conservatory and several greenhouses. Each of these charges an entrance fee. The agency operates lucrative car parks along the Niagara Parkway. Those who wish to park close to the brink of the Canadian Falls can do so for a high price, one that is similar to a downtown car park in a major city. Those who wish to pay less for parking can use cheaper car parks further upstream, and then take an agency transit bus to the Falls. The agency operates a series of restaurants along the Parkway, ranging from fast food to fine dining. All the major souvenir stores along the Parkway are owned and operated by park agency personnel. The highly successful Maid of the Mist travel boats that take tourists to the base of the raging torrents are operated under licence by a concessionaire, with a financial return to the agency. The agency provides landscaping services for a fee to the massive hydro-electric facilities along the Niagara River. The programme is designed to mask the visual intrusion of these facilities within the parkland. The commission even operates its own currency exchange facilities and an automated teller machine. The Parks Commission maintains that it services up to 16,000,000 visitors a year, a figure that is a rough estimate. Nevertheless, the huge volume of use, combined with the income from a wide variety of tourist services, provides a very successful programme, both financially and touristically.

The fiscal success of this agency is due to its structural ability to function like a corporation. It has a monopoly within its lands. It makes its own decisions on service provision. It can set it own fees. It can retain all revenue and use this to cover costs and finance new operations. It is important to note that this operation is within the context of its public mandate to provide a cost-efficient tourism experience for domestic and international tourism along the Niagara Falls and River.

It is important to note that within a park system, not all parks can generate sufficient revenue to cover operation. Low-use nature reserves, for example, may have management costs that are much in excess of the site's capability to support tourism. Some high-use parks can produce a surplus that can be transferred to other low-use parks in a system. Agency-wide efforts in cross-marketing and intellectual property licensing can provide substantial income. These income pools can be used to subsidize the low-use site. It is important to balance the needs of any one site and its internal fiscal ability. At least a portion of fee revenues should be earmarked for the site that generates the income. Earmarking increases site management's incentives to collect fees efficiently.

It is important that visitors are made aware of where their fee payments are going. They are often willing to pay appropriate fees when they know that fees are used for services and programmes that they appreciate. Fee systems must be supported by reliable accounting and management. Standard business practices for fee collection, employee verification and income tabulation are necessary.

Those park organizations that are government agencies often lack the structural abilities to function properly as businesses. They have complicated and slow procedures to set prices. They cannot go to capital markets for loans. They cannot retain revenue; it must all be sent to central government.

Given these limitations with government agencies, many governments are restructuring their park agencies into parastatals, or corporations within government. The Kenya Wildlife Service, the Tanzania National Park Agency and the South African National Park Agency all function as parastatals, with the financial advantages visible in the Niagara Parks Commission example given above.

A few park agencies are experimenting with the licensing of intellectual property. The names and images of national parks and game reserves are some of the most well-known and powerful in the world. Serengeti National Park, Amboseli National Park, Kruger National Park and Banff National Park are some of the most famous brand identities in the world. Private corporations will often pay high sums for the use of these names and images. This may be done in unique ways, such as with charges associated with movie filming on location in parks.

Cross-marketing occurs when one product or organization advertises in concert with another. An example could be a park agency using one type of recreational vehicle, thereby advertising to all the visitors its special qualities in the park environment. In concert, the vehicle manufacturer would publicize the park as the point is made about the special features of a vehicle. Such arrangements are now being developed.

Australia is a typical example; most park agencies in the country rely on only a few of these sources of income, typically, entrance fees, some recreation service fees and accommodation fees, usually for camping (Queensland Department of the Environment, 2000). Australia has a long tradition of free public access to natural and cultural heritage assets, so much so that when the Great Barrier Reef National Marine Park proposed an increase from Aus$1 to 6 for park visitors using commercial tourist operators, a Senate parliamentary committee inquiry was launched (Allison, 1998). The percentage increase was large, but the absolute charge after the increase was still less than the cost for entry to almost any other leisure site. This inquiry came to the apparently self-evident conclusion that: 'It must be accepted that user charges can usually raise no more than a small percentage of total costs' (Allison, 1998, p. 133). Clearly, the Australian Senate did not do their fiscal research adequately. This inquiry apparently did not recognize, as is commonly the case, that there are many sources of income available from various tourism sources, not just entrance fees. It is not true that parks are restricted in income to only a small percentage of total costs, as was shown with the Niagara Parks Commission example.

In several countries dramatic increases in park use fees were introduced without proper client consultation, most specifically in Costa Rica and Zimbabwe, resulting in vociferous objection and subsequent rollback of some of the increase. The lack of knowledge of pricing policy and the methods of price adjustment is common in parks, and was evident in these two countries.

Concession Policy

Private companies often provide services, programmes and facility operations within parks (Fig. 12.4). Typically, they operate under the conditions of a licence obtained from the park. For example, a park may

Fig. 12.4. Private companies sometimes provide services, programmes and facility operations within parks. The Tendele Hutted Camp provides valuable income for Royal Natal National Park and for KwaZulu-Natal Wildlife Service. This park service found that it can earn more income by operating the Tendele Camp directly, rather than leasing it to a private company. Dawn at Tendele Hutted Camp, Royal Natal National Park, South Africa. (Photographed by Paul F.J. Eagles.)

require all tour guides to have specialized education and to carry liability insurance in order to gain a guiding licence. There may be no limit to the number of licences given, or there may be a finite number of licences issued. The licence may be exclusive, with no other similar licenced operation permitted. This is often called a concession. A typical concession would be that of a private company operating a restaurant and a food store in a park building.

Virtually every park agency has a policy for fees, licences and permits. The management of concessions is one of the most challenging and difficult aspects of park management.

Concessions are utilized by park agencies for several reasons. Many park agencies do not have the legal structure to function with the management efficiency and effectiveness necessary to operate a business. For example, they may not be able to retain revenue from sales because all revenue must go to central government. Other examples include lack of ability to access private capital, inflexibility in pricing policy and rigid labour contracts. In these situations, private companies are used to provide specialized services and products. Such companies can quickly adapt to the market with innovation and rapid response to demand changes. They have access to private capital. There may be much more flexibility in labour contracts, compared with government.

The private sector has the ability to act promptly to offer a service when there is the possibility of a profit. The private sector will only operate programmes that provide strong financial return. They may not want to trade in low-volume periods or provide services at average prices. Some services may not be profitable to the private sector and these will simply not be offered or park management may have to subsidize the operation.

The goal of a concession, from the public's point of view, is to further the purposes of the park, to provide access to the heritage resources compatible with the legislation, and to provide for the needs of visitors. The contract must detail the services required, their timing and their quality. Concessionaires operate within a special, sensitive, natural and cultural environment. It is necessary that their staff members are suitably trained for operation in such an environment. There are many operational details, such as hours of operation, range of services, level of service and employee qualifications that must be outlined in the contract. A fundamental issue is that of pricing policy. When a park concession has a monopoly, regulation of prices may be desirable. When there are many operators, competition is encouraged.

With regard to contract length there is potential conflict between the needs of the concessionaire and the needs of the park. Concessionaires prefer a longer-length licence period in order to establish the business, to earn sufficient return on initial capital expenditures and to gain maximum profits. The higher the initial costs with the establishment of the contract, the longer the desirable period of the contract. The length of the contract must be long enough so that the company has time to develop their procedures, explore the market and establish a solid business presence. Park managers often prefer a shorter tenure in order to maintain flexibility. For example, a long contract may lead to complacency. A term of from 5 to 10 years is often chosen. Park managers should undertake annual monitoring and evaluation of the contract performance.

Often, the basic facilities, such as the store or the campground, are owned by the park but are leased to the private sector for a period of years. Sometimes the infrastructure is constructed by the concessionaire, but is transferred to the park after a period stipulated in the contract. A lease/lease-back procedure is sometimes used to gain tax advantages. In this case, the infrastructure is built by the concessionaire, transferred to the park on completion and then leased back to the concessionaire. Occasionally, the tourism facilities are owned by private enterprise under a permanent land lease. This can be disadvantageous to park management because of the inability of the park to manage activities and behaviour of privately-owned facilities in a park. The park agency has little ability to influence a private facility's policies and operations if the private facility has permanent rights in a park.

The concession or licence contract outlines the rights and responsibilities of each party. Issues covered in the contract are listed in Table 12.7.

Table 12.7. Concession contract issues.

Hours of operation
Customer service standards
Environmental practices
Pricing policy
Public access to facilities
Infrastructure maintenance responsibilities
Signage
Advertising
Standards for operations and staff
Facility design and maintenance

It is important that the responsibilities of each partner, the concessionaire and the park are listed in sufficient detail. It is useful to measure performance of the contract at periodic intervals. Penalties for non-compliance must be clearly stated. There must be a procedure outlining the rules for cancellation of the contract due to non-compliance with the contract.

Typically, the park receives a fee set in the concessionaire contract. This fee can be calculated in many ways: it can be an annual fee at a pre-set level, an annual fee plus a percentage of gross revenue; or a percentage of gross revenue only. The fee payable can be gradually increased over time to encourage growth and innovation. The fee can be structured to provide encouragement for the concessionaire to operate at a specific time, for example a lower fee in low-volume periods.

The choice of the concession company is critical. The selection procedure can be highly political, with abundant prospect of political interference. It is an occasion for park staff to undertake self-serving behaviour, such as demanding a bribe or being financially connected to the contractor. Selection procedures must be fair to all parties, open, transparent and neutral. Independent tendering of all contracts is necessary.

Concession management is one of the biggest problems for park managers. Concessionaires sometimes ignore contract conditions. They have been known to illegally construct facilities on parkland or operate a business not allowed in their contract. Their employees can be of low quality and cause considerable trouble, such as theft and environmental damage. Concessionaires often try to avoid contract rules by petitioning senior government officials or influential politicians. Private operators may show little desire to assist with other aspects of park operations, such as provision of accurate information, assisting injured visitors or helping in emergency situations. Once a bad operator gets into place, it can be very difficult to get him or her removed. The enforcement of concession contracts and the policing of concessionaires can be very expensive and time consuming for park managers.

In recent decades NGOs have become more prominent in many parks. These are often called friends groups, such as the Friends of Point Pelee (National Park). A friends group is usually a membership organization, in which local citizens and interested park visitors volunteer money, time and resources. The goals of the group are typically service oriented. For example the Friends of Point Pelee has a goal: ‘to develop and promote activities which contribute to the protection and presentation of the resources of Point Pelee National Park’ (FPP, 2001). Money is raised through the sales of materials, the provision of specialized services and corporate sponsorship. Recently, such NGOs increasingly operate services formerly operated by profit-oriented corporations. This includes lucrative income sources such food sales and equipment rental. In these situations, the NGOs function and are governed in a fashion similar to that of private concessionaires.

The costs and benefits of having park visitor services provided by the park, by NGOs or by private companies must be carefully thought through. There are advantages and disadvantages to each approach. The underlying principles governing the choice of approach should be the provision of high-quality visitor services at suitable costs and within the cultural and environmental constraints of the site. Tourist services can be highly lucrative and it is important for park management to gain the highest possible return, so that other less profitable aspects of management, such as resource management, can be subsidized.

Summary

All parks require sufficient finance (Fig. 12.5). This chapter outlined some of the more important aspects of park finance and marketing and their relationship to tourism. It must not be forgotten that all park operations, especially those dealing with finance, require competent personnel. These people must have the appropriate training. They must be sufficiently rewarded so that they carry out their assigned tasks with diligence. The park must have professional financial systems that ensure that all potential income is collected, recorded and deposited. In the future, most park systems in the world must function fiscally in a responsible and professional manner. Since the income derived from the fees and charges applied to the park visitors is increasingly important as a major source of income, it is critical that all aspects of tourism management are interwoven with the financial aspects of park management.

In the future all park agencies must have highly trained staff in the fields of visitor management, tourism and financial management. The survival of the park and its role in society will increasingly depend on these personnel.

Fig. 12.5. All parks require sufficient finance to cover a wide range of services including public safety. Park visitors often demand access to dangerous sites. Such situations may be very expensive to manage. Monitoring of use, provision of search and rescue services and enforcement of rules are costly. Parks in poorer countries typically are not able to provide the levels of public safety found in parks in wealthy countries. Riverside trail warning sign in Zion National Park, USA. (Photographed by Paul F.J. Eagles.)

References

Allison, L. (1998) *Access to Heritage: User Charges in Museums, Art Galleries and National Parks.* Report of the Senate Environment, Recreation, Communications & the Arts References Committee, Parliament of the Commonwealth of Australia, Canberra.

Burak Jacobson Inc. (1985) *Segmentation Analysis for the Canadian Tourism Attitude and Motivation Study.* Department of Regional Industrial Expansion, Tourism Canada, Ottawa, Ontario.

Eagles, P.F.J. (1992) The travel motivations of Canadian ecotourists. *Journal of Travel Research* XI(2), 3–7.

Filion, F.L., DuWors, E., Boxall, P., Bouchard, P., Reid, R., Gray, P., Bath, A., Jacquemot, A. and Legare, G. (1993) *The Importance of Wildlife to Canadians: Highlights of the 1991 Survey.* Environment Canada, Ottawa.

FPP (Friends of Point Pelee) (2001) *Welcome to our Home Page.* Homepage of the Friends of Point Pelee, online. Available at: www.wincom.net/~fopp/

Hockings, M. (1994) A survey of the tour operator's role in marine park interpretation. *The Journal of Tourism Studies* 5(1), 16–28.

Hockings, M. (2000) *Evaluating Effectiveness: a Framework for Assessing the Management of Protected Areas.* IUCN, Gland, Switzerland.

Langholz, J. (1996) Economics, objectives, and success of private nature reserves in Sub-Saharan Africa and Latin America. *Conservation Biology* 10(1), 271–280.

Lindberg, K. and Huber, R.M. (1993) Economic issues in ecotourism management. In: Lindberg, K., Hawkins, D.E.,

Epler Wood, M. and Engledrum, D. (eds) *Ecotourism: a Guide for Planners and Managers.* The Ecotourism Society, North Bennington, Vermont, pp. 103–104.

Queensland Department of the Environment (2000) *Benchmarking and Best Practice Program: User Pays Revenue.* ANZECC Working Group on National Parks and Protected Areas Management. Queensland National Parks and Wildlife Service, Brisbane, Queensland.

Thomas, L., Curtis, R., Dixon, J., Hughes, G., Sheppard, D., Rosabal, P., Vorhies, F. and Bagri, A. (2000) *Financing Protected Areas.* IUCN, Gland, Switzerland.

CHAPTER 13
Park Tourism Policy

Introduction

The management of a park's visitation is a vitally important component of the park's profile in society. The involvement of key stakeholders in the development and implementation of the policy is vital. The experiences of visitors, their satisfaction with their visit and their comments to others partially determine a park's identity (Fig. 13.1). The role of the local community, the level of involvement by local tourism businesses and the amount of impact on local people's lives influence the local community's identification with the park. Therefore, park tourism policy is a critically important aspect of park planning and management. This chapter discusses key issues in the development of park tourism policy.

National parks, game reserves and other forms of protected areas can provide highly valued tourist destinations. Table 13.1 shows the reported levels of use for several individual parks highlighted in this book. These parks show the very wide range of visitation levels that occur, ranging from Aulavik in the Canadian high arctic with 63 visitors in 2000 to Great Smoky Mountains National Park in the mountains of the USA with 21,110,495 entrants in 2000. This range of volume and type of visitation is very challenging for managers.

Great Smoky Mountains National Park may have the largest park visitation in the world. The 21,110,495 entrances consist of 10,175,812 recreational visitors and a further 10,934,683 entrants that were not recreational. This park contains a major highway. This highway accommodates a large amount of traffic that is just passing through. Therefore, the park managers have responsibility for managing both a very large recreational use and a very large travel use that is not related to the park and its natural resources. It is necessary for park managers to collect and report information on both entrants and visitors to parks. It is equally important for users of these data to understand the differences between recreation visits and

Fig. 13.1. The experience of the visitors, their satisfaction with their visit, and their comments to others partially determine a park's identity. The Niagara Park Commission, Canada's oldest park agency, provides safe and easy access to the dramatic edge of the mighty Canadian Niagara Falls. High levels of public satisfaction with this experience encourage further visitation and positive public expressions of satisfaction. Niagara Falls, Canada. (Photographed by Paul F.J. Eagles.)

Table 13.1. Visitation levels for selected parks.

Park	Year	Visitor days
Yellowstone National Park (USA)	2000[b]	3,794,703
Great Smoky National Park (USA)	2000[a]	21,110,495
Yosemite National Park (USA)	2000[c]	3,550,065
Banff National Park (Canada)	2000[d]	4,635,705
Aulavik National Park (Canada)	2000[d]	63
Lake District National Park (England)	2000[e]	~12,000,000
Tsavo National Park (Kenya)	1996[f]	231,100
Volcan Poas National Park (Costa Rica)	1999[g]	226,736
Kruger National Park (South Africa)	2000[h]	804,069

[a]Parknet (2001a); [b]Parknet (2001b); [c]Parknet (2001c); [d]Parks Canada (2001); [e]Lake District National Park Authority (2001); [f]Kenya Wildlife Service (1997); [g]Baez (2001); [h]Cupido (2001a).

entrants. The high use level in the Great Smoky Mountains National Park requires a very large and sophisticated public use management system. The impact of such a large volume of tourism on local communities is immense. Local communities, regional communities, local states and distant cities are all affected by the use and are interested in all aspects of the park's tourism management policies.

Many parks have heavy usage in the range of 3 to 5 million visits a year (Fig. 13.2). Table 13.1 lists Yellowstone, Yosemite and Banff in that category. As a result of this use these three national parks have high national identities and significant international profiles. These three parks are so well known globally that they might be considered to be international parks, not just national parks.

Kruger National Park in South Africa caters for 0.8 to 0.9 million visitors a year, 25% of which are foreign. Tsavo in Kenya and Volcan Poas in Costa Rica are both the most highly used national parks in their respective countries, each with around 225,000 visitors a year. Even at these more modest levels these parks are the anchors of very important nature-tourism industries in their respective countries. The foreign visitation is particularly important because of the foreign currency earned.

It is important to note that many national parks and protected areas have very low visitation. For example, the Province of Ontario in Canada had 278 provincial parks in 2002 covering 7.1 Mha. Of that number, 174,

Fig. 13.2. The Niagara Parks Commission provides facilities capable of handling large numbers of tourists who view Niagara Falls up close. This photo shows the area at the lip of the Canadian Falls, which services over 16,000,000 people a year. The service building holds a restaurant, stores, ticket booth, toilets, telephones and a money exchange. Virtually all the tourism services along the Niagara Falls and Gorge area in Canada are operated directly by the park agency. Mass nature tourism, Niagara Falls, Canada. (Photographed by Paul F.J. Eagles.)

covering 4 Mha, had no reported visitation (Mulrooney, 2001). Typically, these non-operating parks had no on-the-ground personnel, as well as very few facilities and programmes. Parks with very low visitation have a weak public profile, resulting in low tourism demand. Such low demand results in low government investment in the park and its facilities.

However, reported visitation figures can be misleading. Some parks that report no visitation actually have use, which may be substantial. The lack of reporting is due to insufficient finance and management capability in the parks. Parks with existing tourism use but very weak management are candidates for negative impacts. Parks that report no visitor use will be undervalued by government and the local community. They will be in a weak position to argue for appropriate levels of government finance to undertake adequate management.

Individual parks exist within a system of parks, such as a state park system or a national park system. The impacts of the total usage of the entire system are of a different scale from that of individual parks. Table 13.2 shows the use levels for selected park systems, chosen because of the discussion of these systems in this book. Clearly, some parks and some park systems have notable levels of visitation and therefore have significant tourism industries.

The USA National Park System catered to 429,853,123 visits in 2000. This makes it the most heavily used park system in the world, and one of the largest tourism enterprises in the world. Appendix A contains the operational policies for tourism by this agency. Given the immense size of tourism involving USA national parks, every component of this policy is of importance to a large number of people, businesses and communities. The USA also has a large and prominent system of National Wildlife Refuges. This system caters to a much more modest, but still quite impressive, 27,700,00 visits. Parks Canada serviced 15,690,073 visits to the national parks in 2000, and an additional 9,737,051 visits to national

Table 13.2. Visitor use levels of selected park systems.

Park system	Year	Visitation levels
USA National Park System	2000[a]	429,853,123
USA National Wildlife Refuge System	1995[b]	27,700,000
Costa Rica National Park System	1999[c]	866,083
Canada National Park System	2000[d]	15,690,073
British Columbia Provincial Park System (Canada)	2000[e]	24,271,044
California State Park System (USA)	2000[f]	~70,000,000
New South Wales Park System (Australia)	1997[g]	19,988,281
Kenya National Park System	1996[h]	1,381,700
South Africa National Park System	2000/2001[i]	2,440,902

[a]Parknet (2001d); [b]Laughland and Caudhill (1997); [c]Baez (2001); [d]Parks Canada (2001); [e]British Columbia Parks (2000); [f]Areias (2001); [g]A. Ramsey (2001, personal communication); [h]Kenya Wildlife Service (1997); [i]Cupido (2001b)

historic parks. Parks Canada operates two distinct park systems, one of national parks and one of national historic parks and sites. The use data is therefore reported in two categories. The National Park Service of the USA merges the historic and natural park use data into one category.

Kenya, South Africa and Costa Rica have significant nature-tourism industries based on visitation to their national park systems. Kenya serviced 1.4 million national park visitors in 1996. South Africa had 2.4 million in 2000/2001 and Costa Rica 0.9 million in 1999. South Africa's park visitation was 16% foreign and Costa Rica's 43% foreign.

In terms of geographical scale the national park systems of Kenya and Costa Rica might be better compared to State Park or Provincial Park systems of the large federated countries. California State Parks in the USA service an estimated 70,000,000 visits a year, making this system the most heavily visited state park system in the USA and one of the most heavily visited systems in the world. In Canada the British Columbia Provincial Park system is the most heavily visited at 24.2 million visits in 2000. In Australia, New South Wales's State Parks cater to approximately 20 million visits a year.

In some countries, such as Canada, national park tourism stayed relatively stable over time (Fig. 13.3). However, in the same period the provincial park visitation increased, suggesting that in this country the provincial parks were more attractive, and possibly more proactive, for park visitation than national parks. Figure 13.3 shows that overall figures include visitation for both national parks and national historic parks. The historic park use constituted approximately 40% of all national park use.

In many countries park tourism has increased over time. Figure 13.4 shows the increase in the visitation to Costa Rican National Parks of over 400% in a 15-year period. During this time, Costa Rica developed into a

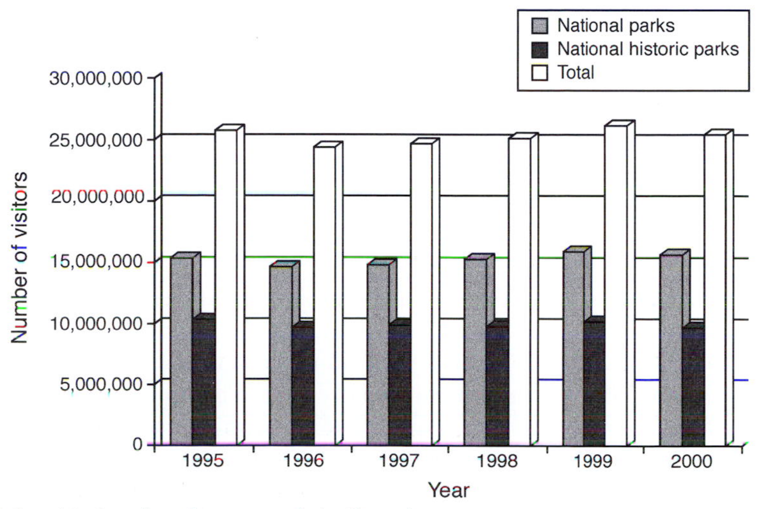

Fig. 13.3. National park use trends in Canada.

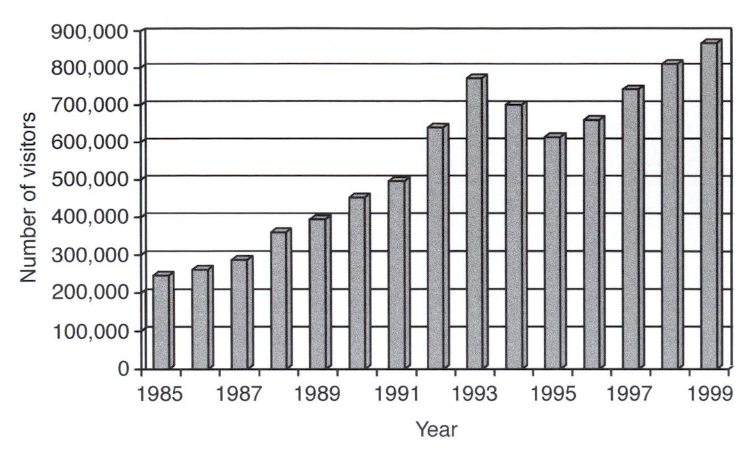

Fig. 13.4. National park use trends in Costa Rica.

premier ecotourism and park tourism destination (Eagles and Higgins, 1998). The steady increase was halted and briefly reduced when the national park agency introduced a surprise 800% increase in fees, combined with a reduction in international travel due to an economic downturn. With more appropriate pricing policy and a better international economy, the increase resumed.

Clearly, park tourism is a major tourism enterprise in many areas. Therefore, all aspects of park tourism policy have major influence and must be carefully developed, implemented and monitored. Foreign visitation is particularly important because of its contribution to the economy as an export, that is an activity that brings foreign currency into the country.

The Roles of Public and Private Enterprise in Park Tourism

In parks with higher levels of visitation there is usually a complicated division of labour and responsibility between the public and private sectors in servicing park visitation. In Chapter 12 we discussed the typical division of responsibilities between the public and private sectors in serving park visitors. The public sector has the major responsibility for: the environmental protection of resources; the construction and maintenance of the infrastructure (roads, airports, rail lines, electricity and sanitation); visitor security and rule enforcement; the monitoring of social and environmental impacts; the evaluation of environmental and service quality; the allocation of access to resources; the determination and enforcement of limits of acceptable change; some information provision for the visitors (interpretation, visitor centres); the resolution of conflict; some guiding and recreational programmes; and some accommodation (typically campgrounds but sometimes hotels and cabins). These responsibilities are carried out within the park boundaries. The private sector has chief responsibility for some accommodation (typically the larger, more expensive forms such as lodges and hotels); the provision of food to the visitors (such as in restaurants and in food stores); the transport of the visitors and servicing of the visitors' equipment (buses, automobiles, boats and aeroplanes); some level of information provision (guides and interpretive programmes); most of the advanced media (films, books and videos); some of the park promotion, most of the consumer products (clothes, souvenirs and equipment); and nearly all of the personal services (entertainment). The private sector tends to provide these services both outside the park and inside the park.

A typical example of the interweaving of the division of labour in parks is that of transport. The public sector typically provides the infrastructure, such as the roads, car parks and airstrips, while the private sector provides the transport equipment that uses the infrastructure, such as the cars, buses and aeroplanes. The public sector determines the allowable levels and timing of use of this infrastructure. The private sector builds and services all the transport equipment. This intricate division of labour means that there must be coordinated action by both sectors for the fulfilment of the transport needs of the park visitors. An equally important aspect, discussed in Chapter 12, is the division of costs and benefits in such an integrated enterprise. A complicated set of grants, taxes, fees and charges are necessary to obtain the funds necessary to operate the system. In wealthy countries most of the capital costs for transport infrastructure are derived from government taxes on society, while the operating costs of the infrastructure are derived from both taxes and fees. In poorer countries, the capital costs are often covered by both foreign aid grants and tourist fees. Typically, in these situations all of the operating costs must come from the tourists because there is no alternative source of funds. Either the tourists pay for the costs, or the facilities and services will not be available for tourist use.

This division of responsibilities, costs and benefits developed over long periods of negotiation and debate. Each party attempted to maximize its benefits, while limiting its costs. For example, it is common for the private sector to complain about government taxes, while at the same time making strong demands for the expenditure of tax funds on desired facilities. It is common for the negotiations to take place simply between directly interested parties, as in the case of the park and private transport companies. In this situation, the larger implications of the decisions for other parties and for the environment may not be fully dealt with.

Given this complicated interwoven situation of public and private responsibilities, it becomes obvious that an equally complicated set of sophisticated tourism policies governing these responsibilities is required. The development of these policies involves long and detailed negotiations among all the principal parties, such as the park agency, relevant private tourism corporations, the park visitors, the local communities and special interest groups. It is clear from the example of tourist transport why the use of a comprehensive public participation programme in the development of all aspects of a park tourism policy is essential.

Tourism Management Approaches and Models

Policy is a statement of intent. A written agency policy is a statement that guides the actions of the agency staff. It also provides information for all stakeholders, public and private, who are interested. Typically, government park agencies follow written policy that outlines the goals for park visitation within a park system and within a specific park. However, many agencies have unwritten policy, day-to-day actions that, over time, develop a history and a substance. Often this unwritten policy is as important to the management of park visitation as the published policy. This situation of having both a written and an unwritten policy makes it very difficult for the interested observer to gain a full understanding of the details of park tourism management in most parks.

Most park agencies have a park visitation or a park tourism goal. These goals can vary from simply allowing certain types of visitation, to operating sophisticated visitor management programmes with volume and quality goals. Only a few agencies have detailed tourism planning and management policies. All park agencies do some tourism planning and management, but often in an *ad hoc* fashion, following the immediate needs without the guidance of an overall policy structure.

In Canada the Ontario Provincial Parks system has an approved policy goal. It is the goal of this park system: 'To ensure that Ontario's provincial parks protect significant natural, cultural, and recreational environments, while providing ample opportunities for visitors to participate in recreational activities.' More specific statements on four areas refine this overall goal: tourism, protection, recreation and heritage appreciation.

The government of Ontario's approved tourism goal for all Ontario Provincial Parks is 'to provide Ontario's residents and out-of-province visitors with opportunities to discover and experience the distinctive regions of the province'. This simple statement indicates that these parks have a role in providing recreation opportunities to provincial residents and more universally to other Canadians and to foreigners. Therefore, park managers have the responsibility to provide recreational facilities and programmes that service a broad tourist market. It also implies that parks will be located in all regions of the province. This suggests an overall system plan must be developed that looks at park and tourism objectives for all geographical regions of this large province. The tourism objective is balanced by the other three park system objectives for protection, recreation and heritage appreciation. It is important to recognize that the application of any one objective is balanced by the goals inherent in other objectives. This balancing between multiple objectives, some of which may be in conflict, is a critical component of policy development and application.

Two examples of comprehensive park tourism policies are discussed in this chapter. The National Park Service of the USA is the largest park tourism provider in the world. Given the complexity of such an operation, the published policy serves as an important indicator of the issues involved. The New South Wales National Parks and Wildlife Service manages the national parks of the State of New South Wales in Australia. This agency developed a comprehensive park tourism policy in the latter half of the 1990s. This policy has many innovative features that illustrate the range of issues emerging in recent years in park tourism policy.

The tourism policy of the National Park Service of the USA

The National Park Service of the USA is one the few park agencies with an approved tourism policy (see Appendix A to this chapter). This policy provides direction on the types of issues to be addressed by a park agency in tourism. Actions include the following:

1. Dialogue and outreach with other public and private tourism interests.
2. Show agency leadership in sustainable tourism design and operation.
3. Highlight national diversity.
4. Encourage visitation by peoples of all types.
5. Provision of cost-effective and accurate information services.
6. Encourage visitation of low-use parks, and off-season use of high-use parks.
7. Management for international visitation.
8. Identify desired resource conditions and visitor experiences and develop procedures to provide these conditions.
9. Influence the plans of tour operators and gateway communities towards park goals.
10. Mediate the relationships between park concessionaires and other aspects of tourism services.
11. Keep the agency up to date on tourism trends.
12. Look for funding partners to help carry out park programmes.
13. Keep key stakeholders, such as local communities and private tourism businesses, informed about resource conditions, resource management and safety issues.

These issues reveal the types of societal pressure placed on the national parks of the USA. They must show sustainable tourism leadership, while encouraging cost-effective tourism usage by a wide diversity of people. Policies must be developed for cooperation with a wide range of other public and private tourism interests, including those companies that have special contractual arrangements to provide tourism services in the parks. In Chapter 12 we discussed some of the details involved in concessionaire policy (Fig. 13.5). The National Park Service must have an ongoing information programme for both park visitors and key stakeholders, such as local communities and private tourism businesses. The agency must try to shift some visitation from peak periods to shoulder seasons. Objectives for natural and resource conditions must be developed in concert with interested stakeholders. Visitor-experience objectives must be developed and implemented. Clearly, national parks in the USA have many challenging and possibly conflicting tourism objectives that must be implemented.

It is a cardinal rule of policy development that goals must be worded so that their implementation can be properly assessed. It is useful to look at one policy that may be problematic in this regard. Policy 4.3, in Appendix A, aims to: 'Encourage practices that highlight America's diversity and welcome park visitation by people of all cultural and ethnic backgrounds, ages, physical abilities, and economic and educational means.' The actions are encouraged and welcome. It should be fairly easy to assess whether the agency has achieved encouragement of diversity and welcoming of visitation. However, there are no concrete, quantitative goals, such as a specific distribution of use of identified socio-economic groups. Many of the NPS goals are written in such a way that assessment of policy implementation is problematic.

Fig. 13.5. Park policy outlines the relative amount of public and private provision of services to park visitors. This tour boat is operated by a concessionaire to the Niagara Parks Commission. A private company provides the tourism service under contract, with a financial return from tourism earnings going to the park agency. Park agencies typically utilize concessionaires to provide specialized services that require unique equipment, training and business arrangements. Maid of the Mist Tours, Niagara Falls, Canada. (Photographed by Paul F.J. Eagles.)

Other goals will be easier to evaluate. Goal 4.6 encourages the NPS to: 'Pursue practices, such as the use of universal design and the inclusion of metric measures on signs and printed media, that will contribute to the safety and friendly accommodation of all visitors.' The inclusion of metric measures on all signs and printed media is an easy goal to evaluate. However, universal design is not defined and very much open to debate. Specific rules in regards to visitor safety must be developed. Similar work is necessary to assure that the signs and media are user friendly.

It is important to consider if the wording of these policies allows the monitoring of implementation. There is no mention in the National Park Service policy that monitoring will occur, or that this monitoring will result in a determination of policy implementation success or failure.

The NPS tourism policy appears to be a very general statement of intent, one that provides general direction. However, it is vaguely written and often difficult to evaluate.

The tourism policy of the New South Wales National Parks and Wildlife Service

In 1997, the New South Wales National Parks and Wildlife Service released a *Draft Nature Tourism and Recreation Strategy* (NSWNPWS, 1997). This is the most detailed park tourism strategy written, and it is worthy of discussion. Appendix B provides a list of the principles that guide this strategy. These principles are basic assumptions that are non-negotiable. All policies must develop from and take account of these principles (see Fig. 13.6). For example, the paramount purpose of the

Fig. 13.6. The Penguin Island Discovery Centre is a penguin rehabilitation site developed as an interpretive centre for interested naturalists. The combination of wildlife biology, interpretation and tourism is a popular mix of activities. Discovery Centre, Penguin Island Reserve, Western Australia. (Photographed by Paul F.J. Eagles.)

parks is conservation of natural and cultural heritage and all management must aim to meet this objective. The statement of fundamental concepts underlying policy and management is an excellent approach. It provides a foundation on which all policy is built. It helps to explain the context for policy. It outlines fundamental truths of policy that may be difficult and not avoidable. For example, all decisions have a cost and sometimes actions may have to be taken without full scientific information.

There are four desired outcomes of the strategy. These are:

1. Ecologically sustainable visitor use of the protected areas of New South Wales (this is the primary outcome and next three are secondary to this one).
2. Positive assistance provided to facilitate sustainable nature tourism and recreation in New South Wales.
3. Positive assistance provided to facilitate improved economic return to New South Wales through sustainable nature tourism and recreation.
4. The improvement of the heritage conservation condition of protected areas in New South Wales.

The overall objective is ecologically sustainable visitor use. The secondary objectives are the development of a sustainable nature tourism industry in the state and improvement of the condition of the parks (Fig. 13.7). Presumably, the policy makers are anticipating that a fully functional nature tourism industry will be financially self-sufficient and will assist parks in their financial needs. The New South Wales Policy is innovative in this regard. Most often park tourism policies ignore their financial implications. This is very strange since park tourism is probably the most expensive activity for the park to manage. Conversely, park tourism is often potentially the most lucrative activity for the park. This lack of proper business goals in regard to the financial costs and benefits

Fig. 13.7. Little blue penguins nest along the coast of Southern Australia. Each year many are injured or orphaned and are given to government authorities for rehabilitation. Wildlife tourism is a rapidly developing activity in Australia. Little blue penguin, Penguin Island Reserve, Western Australia. (Photographed by Paul F.J. Eagles.)

of managing visitor use of parks is a major reason for the failure of many park management plans. It is also the major reason that many park tourism efforts are beset by the problem of too high a level of negative social and environmental impact. Due to a lack of proper financial planning, the parks do not have the money to properly manage the tourism flows. Without sufficient financial and human resources, every management programme has a high chance of failure. Therefore, it is noteworthy that New South Wales recognizes financial sustainability as an overall objective.

The fulfilment of these four objectives in the New South Wales Draft Policy involves action on many fronts. It is worthwhile to note some of the key action areas outlined in the draft strategy.

1. Each park shall have nature tourism and recreation plan as part of the Plan of Management.
2. Clear and precise ecological, environmental, economic and managerial planning objectives shall be developed for each park. A recommended range of visitor use numbers will flow from these objectives, taking into account the ability of the park to sustain this use level.
3. Commercial tour operator use levels will be estimated. When there is a visitor use limit, a proportion of the use will be assigned to commercial operators.
4. Research sites on visitor use and ecological impacts will be established.
5. Monitoring of guideline implementation will be done.
6. A statewide recreation setting inventory will be done, using the recreation opportunity spectrum (ROS) six-class system as a basis.
7. Management guidelines will be developed for each ROS class.
8. All inventories and classes will be represented with a geographical information system.
9. New agency staff expertise will be developed in nature tourism, especially in economic evaluation, social impact evaluation and ecological conservation.
10. The staff performance appraisal system will be amended to include visitor use management performance.
11. A new statewide non-commercial licensing permit system will be investigated.
12. Active encouragement of research into ecologically sustainable visitor use will be encouraged.
13. New and more accurate systems for visitor use monitoring will be developed.
14. Opportunities for cooperation will be investigated with the State Forest System and with Aboriginal Boards of Management.
15. Community involvement in the development of tourism plans will be encouraged.
16. Investigation of commercial operator licensing systems will be done.

These recommendations give a sense of the depth of policy development and implementation that is necessary for the full implementation of tourism management in parks. Key issues discussed that are seldom mentioned in other tourism policies include: a statewide recreation setting inventory, new agency staff expertise to be developed, a new staff performance appraisal system and a new licensing permit system to be undertaken. These policy statements are comprehensive and have a high opportunity for success in implementation.

It is important to note that the Draft NSWPWS tourism policies spurred an active public debate on the costs and benefits of park tourism. One key universal issue was the acceptable level of visitation. How much park visitation is desirable? Another key issue was the source of park management funds. Some sectors of society felt that most if not all park funding must come from government tax revenues, and that the operation of a park like a self-sustaining business is improper. It is a very common position for user and environmental groups to demand that society should cover all the costs of management, with the users' programmes and facilities subsidized by taxpayers. A critical point in this discussion was the relative amount of costs to be covered by government grants or by tourism fees and charges. This is a universal debate that has quite different outcomes in different countries.

All over the world governments are demanding that the users of parks pay a higher proportion of the costs of park management. The concept is that those who benefit most directly should pay the costs. This is an underlying trend visible in New South Wales at this time.

Most of the policies require more detailed development. For example, the statement directing a new staff performance appraisal system will require negotiations between managers and staff with regard to the methods to be used to implement the proposed visitor use management performance. Such a policy development will involve sophisticated understanding of service quality measurement, of staff reward systems and of labour law. It is to be expected that labour negotiations will be required between the staff union and the agency management. Each of the overall policies starts off a cascade of further actions, such as these.

The New South Wales Draft Park Tourism Policy is written in such a way that most of the goals can be evaluated towards success or failure. This is one of the most advanced statements of its type. It is an innovative approach that is more comprehensive than most park tourism policy statements.

Summary

Older park systems have significant experience in the operation of park tourism. Much of this was developed through trial and error management over time. It is important to use this existing knowledge background in the development of new policy statements. However, many parks manage tourism in an *ad hoc* fashion, basing their actions on what worked yesterday. This approach is viable in the short term, but can lead to considerable problems in the longer term. Therefore, it is essential that sophisticated, long-term policies for park tourism are written and exposed to open public debate at the earliest possible opportunity.

Park tourism policy development and implementation require sophisticated planning and management procedures. It is best done with specially trained personnel with experience in policy development, tourism planning, public participation and park management. The success of park tourism management depends on a well-coordinated set of policies that are accepted by the majority of the involved stakeholders. All policies must be monitored to evaluate them for success or failure. Failures must be corrected (Fig. 13.8).

Park tourism policy is a critical component of park management. It deserves a high degree of attention in all park policies and park management plans.

Fig. 13.8. Park policy determines the type and location of waste disposal. Policy also determines the housing of park staff and residents. This photo shows one result of a housing and waste-disposal policy. Poor waste handling creates a magnet for wildlife, such as bears and deer. Over time park agencies have become much more sophisticated in their waste management. Garbage and deer, Yosemite National Park, USA. (Photographed by Paul F.J. Eagles.)

References

Areias, R. (2001) *The Seventh Generation*. California State Parks, Sacramento, California.

Baez, A.L. (2001) Costa Rica Como Destino Turistico. Unpublished Conference Paper, Rio De Janeiro, Brazil.

British Columbia Parks (2000) *BC Parks*. Annual Information Exchange. Federal Provincial Parks Council, Thunder Bay, Ontario.

Cupido, R. (2001a) Kruger National Park Tourism Performance March 2001. Unpublished data, South Africa National Parks, Pretoria, South Africa.

Cupido, R. (2001b) SA National Park Tourism Performance March 2001. Unpublished data, South Africa National Parks, Pretoria, South Africa.

Eagles, P.F.J. and Higgins, B.R. (1998) Ecotourism market and industry structure. In: Lindberg, K., Epler-Wood, M. and Engeldrum, D. (eds) *Ecotourism: a Guide for Planners and Managers*, 2nd edn. The Ecotourism Society, North Bennington, Vermont, pp. 11–43.

Kenya Wildlife Service (1997) Number of Visitors to National Parks and Game Reserves, 1992–1996. Unpublished internal document. Kenya Wildlife Service, Nairobi, Kenya.

Lake District National Park Authority (2001) *Online*. Homepage of the Lake District National Park Authority, online. Available at: www.lake-district.gov.uk/

Laughland, A. and Caudhill, J. (1997) *Banking on Nature*. US Fish and Wildlife Service, Washington, DC.

Mulrooney, D. (2001) *Provincial Park and Protected Area Statistics*. Parks Ontario, Peterborough, Ontario.

NSWNPWS (1997) *Draft Nature Tourism and Recreation Strategy*. Hurstville, New South Wales.

Parknet (2001a) *Great Smoky Mountains NP*. Homepage of the National Park Service, online. Available at: www2.nature.nps.gov.stats/

Parknet (2001b) *Yellowstone NP*. Homepage of the National Park Service, online. Available at: www2.nature.nps.gov.stats/

Parknet (2001c) *Yosemite NP*. Homepage of the National Park Service, online. Available at: www2.nature.nps.gov.stats/

Parknet (2001d) *National Park System Visitation*. Homepage of the National Park Service, online. Available at: www2.nature.nps.gov.stats/

Parks Canada (2001) *Parks Canada Attendance*. Parks Canada, Hull, Quebec.

Appendix A: Operational Policies for Tourism for the National Park Service of the USA

The 1995 White House Conference on Travel and Tourism established a basis and framework for closer cooperation and mutual understanding between land-managing agencies and the tourism industry. Regional and state tourism conferences have brought park managers and tourism operators together. This dialogue has fostered many of the principles incorporated in the following operational policies.

It is National Park Service policy to:

1. Develop and maintain a constructive dialogue and outreach effort with state tourism and travel offices, and other public and private organizations and businesses, using a variety of strategies, including, but not limited to, memberships in organizations, participation in conferences and symposia, and Internet-based information resources.

2. Collaborate with industry professionals to promote sustainable and informed tourism that incorporates socio-cultural, economic and ecological concerns, and supports long-term preservation of park resources and quality visitor experiences. This collaboration will be used as an opportunity to encourage and showcase environmental leadership by the Service and by the tourism industry, including park concessionaires.

3. Encourage practices that highlight America's diversity and welcome park visitation by people of all cultural and ethnic backgrounds, ages, physical abilities, and economic and educational means.

4. Foster good relationships with park neighbours by promoting visitor and industry understanding of, and sensitivity towards, local cultures, customs and concerns.

5. Provide cost-effective park visitor orientation and information services to visitors in parks and, as funding and partnerships allow, at the visit planning stage, at park gateway communities, and at appropriate threshold locations within park units. As part of this effort, the Service will work to ensure that all who provide information to visitors are well informed and provide accurate information about park activities and resources, including current conditions and seasonal variations.

6. Pursue practices, such as the use of universal design and the inclusion of metric measures on signs and printed media, that will contribute to the safety and friendly accommodation of all visitors.

7. Encourage visitor use of lesser-known parks and under-utilized areas; use during non-peak seasons, days of the week, and times of the day; and visitation to related sites beyond park boundaries, as appropriate, to enhance overall visitor experiences and protection of resources.

8. Specifically address long-term tourism-related trends and issues, and their implications for park plans and management decisions.

9. Represent park needs and realities during the preparation of plans and proposals for gateway community services and park tour operations that could affect park visitation, resources, visitor services and infrastructure support.

10. Promote positive and effective working relationships between park concessionaires and others in the tourism industry to ensure a high quality of service to park visitors.

11. Identify desired resource conditions and visitor experiences, and work to establish supportable, science-based park carrying capacities, as the basis for communicating acceptable levels and types of visitor use, recreation equipment use, tours and services. Carrying capacities are defined for each park as an outcome of the National Park Service planning process.

12. Participate in and monitor travel industry research, data gathering and marketing initiatives to ensure that the Service is fully informed of demographic changes and visitor trends.

13. Work with partners to provide timely, accurate and effective park information, and to ensure that realistic situations and safe, resource-sensitive recreational practices are depicted in promotional materials and advertising. This includes providing appropriate information as early as possible to the tourism industry regarding changes in operations and fees.

14. When feasible, and consistent with park resource protection and budgetary needs, schedule construction, repairs and resource management practices, such as prescribed burns, in ways and at times which keep key visitor attractions and services accessible for public use during peak visitation periods. This will help to minimize adverse impacts on visitors, as well as on park-visitor-dependent businesses.

15. Establish and maintain lines of communication and protocols to handle the impact of park emergencies and temporary closures so that state tourism offices and the public, including tourism communities and tourism-related businesses, have the best and most current information on when park services will be restored.

16. Inform visitors, state tourism offices, gateway communities and tourism-related businesses about current conditions of key park resources and current protection and recovery/restoration measures. Establish a common understanding about what is needed to ensure adequate protection of those resources for present and future enjoyment and how this can contribute to sustainable park-related businesses and economies.

17. Develop new partnerships to help implement Service-wide priorities, and seek partnership opportunities with the industry to fund products and programmes mutually beneficial to accomplish National Park Service mission goals.

Director's Order #17: National Park Service Tourism
Approved: Robert Stanton, Director, National Park Service
Effective date: 28 September 1999
Sunset date: 28 September 2003

Appendix B: Draft Nature Tourism and Recreation Strategy Guiding Principles of Management

The Guiding Principles of Management for the NPWS Nature Tourism and Recreation Strategy are described here.

1. Protected areas of NSW have been reserved for the purposes of conserving a representative sample of the natural and cultural heritage of NSW and the paramount purpose of management is to meet this objective.
2. Visitors are welcome, and visitation to protected areas and the provision of appropriate infrastructure and services will be managed consistently with ecologically sustainable development principles which include:

- the improvement of individual and community well-being by following the path of economic progress that does not impair the well-being of future generations;
- the provision of equity within and between generations;
- protection of biological diversity and the maintenance of ecological processes and systems;
- recognition of the global dimension.

3. Visitor use to protected areas will be managed as ecologically sustainable visitor use.
4. The principles of precaution will be applied when changes to the natural environment are contemplated. These principles are:

- nature is valuable in its own right;
- governments must be willing to take action in advance of full, formal, scientific proof;
- people proposing a change are responsible for demonstrating that the change won't have a negative effect on the environment;
- today's actions are tomorrow's legacy;
- all decisions have a cost (exercising caution may mean some people must forgo opportunities for recreation or profit).

5. State government environmental planning policies and procedures will be followed. Guidelines for both natural and cultural heritage management principles are given in the Australian Natural Heritage and Burra Charters. These principles and guidelines will be followed.
6. Recognition will be given to the special situation of the Aboriginal community, especially their needs for culturally appropriate negotiation and their traditional relationships with the protected areas of NSW.
7. That nature tourism and recreation to protected areas is managed so that it is equitable for this and future generations.
8. The encouragement and regulation of appropriate use involves provision of a range of opportunities for visitors to interact with the natural and cultural features of protected areas, whenever this is compatible with the goal of conserving natural features and processes. The management of a protected area should not, therefore, create artificial features, or promote

the use of specific features in a way that destroys natural and/or cultural values of the park.

9. Recognition that, in evaluating costs/benefits of visitor use, financial analysis alone will not be sufficient.

10. Recognition and maintenance of a wide range of values (including social, cultural, economic, aesthetic and ecological values) in making balanced visitor use management decisions.

11. Recognition that every natural environment has its own special characteristics (sense of place), which must be recognized and respected in managing for visitor use, and in particular when designing and providing (appropriate) visitor facilities.

12. Ensuring, through quality management of destination and support services, that the opportunities for visitors to receive a positive and rewarding experience are maximized.

13. Ensuring that a diversity of recreation settings appropriate for protected areas is available for visitor use.

14. Recognition that protected areas are but one part of a cross-section of land use types within a region, and that an integrated approach to management of visitor use based on strong partnerships is essential to the achievement of sustainable nature tourism and recreation.

15. Ensuring that visitor use of protected areas contributes positively to the natural, social and economic aspects of sustainable tourism.

CHAPTER 14
The Future of Park-based Tourism

Introduction

This textbook is oriented towards improving the stock of knowledge and ultimately the practice of managing tourism in national parks and protected areas. An important component of the job of a protected area manager is not just to deal with the problems of the past and present but also to consider the challenges and opportunities of the future. By considering the future, managers are better prepared to deal with the issues, questions, problems and opportunities brought to them. To a great extent, the future is a function of actions in the present; by anticipating trends, park managers are in a better position to foresee issues and opportunities in the future.

In this chapter, we set forth some of the trends we see occurring that directly bear upon the ability of the manager to practise the stewardship expected by an increasingly sophisticated public (Fig. 14.1). While one may identify other trends, the trends presented here are globally significant ones that must be considered in the practice of national park and protected area stewardship. As a result of these trends, the roles of parks and tourism will expand and evolve in response to changing social needs, roles that will not only help to protect the biodiversity and cultural resources contained within them, but will also provide great opportunity for tourism-related benefits.

Emerging Trends

National parks and protected areas exist within a swirling, dynamic social and political setting that is not only difficult to understand, but is equally challenging to predict. Yet, the setting influences not only the long-term

Fig. 14.1. Low-density wilderness recreation requires very low levels of human use. High demand for such sites results in a lack of balance between supply and demand. Therefore, the allocation of access to wilderness recreation sites is a major park management policy debate that will continue indefinitely. Wilderness canoeing, Killarney Provincial Park, Ontario, Canada. (Photographed by Paul F.J. Eagles.)

planning of parks, but their day-to-day management as well. A growing and ageing world population, increased demands for quality food and higher quality of life, the stability of national institutions, technological change (and conflict with cultural norms and traditions), war and conflict are factors that directly influence the capacity of any nation to adequately protect and manage its natural and cultural heritage. Some of these trends are beyond the capability of park managers to handle. For example, park managers are impotent in the face of an invading army. However, there are trends that, once anticipated, can be successfully handled.

There are a number of trends in political, social, demographic and technological sectors bearing directly upon how and why parks and protected areas are managed. These trends are ones that mark a change in how national parks and protected areas will be managed and influence the roles of those areas in any given culture.

Trend 1. *Growing demands for democratic forms of government translate into increased public participation and collaboration in national park and protected area planning*

One of the interesting phenomena of the late 20th century and early 21st century is increased concern for direct participation in government decision making by citizens. These calls for expanded public participation result from a number of factors including lack of trust between government agencies and affected citizens, the desire for more inclusive and responsive planning processes, a recognition that existing methods of planning often marginalize important values or do not adequately account for the consequences of decisions on citizens, and a general and widespread interest in democratic forms of government. In the western United States, for example, many federal agency decision-making processes now demonstrate a greater level of participation and collaboration than has been practised even in the recent past. This means that the planning processes that protected area management agencies use must be designed to be more inclusive of potentially affected values and interests, provide recognition of the legitimacy of different forms of knowledge, and require planners to acquire facilitation skills. In some cases public demand may cause legislative change in park laws. This change may require parks to involve the public in important park issues, such as the development of the management plan.

Trend 2. *Increased accessibility of sophisticated technology and science in all sectors means that visitors and others interested in parks are more informed and knowledgeable about what opportunities exist and the consequences of various management actions*
The growth of the Internet, in particular, greatly increased the general citizenry's access to information and knowledge that was formerly the exclusive domain of scientists and experts. It also provides inexpensive avenues for many groups to provide information about parks, ranging from tourism companies to environmental groups. This has several consequences for park planning and management. First, it means that potential visitors are more aware of the various opportunities available not only near their home but around the world as well. Potential visitors now have the chance to make plans with greater levels of certainty about conditions and facilities that are available within an area but also are more informed about the alternative destinations from which to choose a cultural or nature-based tourism experience. Secondly, increased accessibility of knowledge means that visitors and others interested in protected area planning issues may provide more informed input into decision-making processes. The widespread availability of knowledge, with a simple click or two of a computer mouse, means that managers will have an increasingly informed clientele with which to interact. In developing nations it is distinctly possible that many park visitors will have access to more sophisticated information than that available to park managers. Thirdly, the widespread availability of the Internet and digital communications means that it is easier for people to communicate across national boundaries and to organize themselves into activist groups promoting one cause or another. Finally, the Internet is an inexpensive method of providing information. Therefore, many groups, such as tourism suppliers, environmental groups and local community groups, can provide abundant levels of information about parks and protected areas. It is difficult for park managers to know what is being said about their park and to ensure that it is accurate and appropriately represents park policy. At present in poorer

countries, third-party interests provide virtually all the park information to tourists. This loss of ability to be the gatekeepers of resource information, policy and management information will have profound impacts on the job facing park managers.

These implications are profound for park managers today and will significantly influence how they do their job tomorrow.

Trend 3. *An ageing population means that there may be significant demand shifts in what activities, settings and experiences visitors seek from national parks and protected areas*

Not only is the world population growing, but also its average age is increasing as scientists make remarkable medical advances in human health. As the population ages, there are potentially significant shifts in demands for recreation opportunities as well as changes in the nature of facilities and programmes required at national parks and protected areas (Fig. 14.2). For example, as people age there is some evidence that they participate more frequently in appreciative and learning activities and less in more active–expressive kinds of activities. And as people age, their needs for supplementary facilities such as wheelchair ramps, trails with lesser grades and other disabled access assistance also rise. In tune with their changing interests, interpretive programmes, particularly those

Fig. 14.2. Private ecolodges cater to those who desire higher levels of personal service. Such lodges will continue to develop in popularity as the ageing population in the developed world produces millions of retirees with abundant time, health, money and a yearning for ecotravel. Marenco is located adjacent to Corovado National Park. Marenco Beach and Rain Forest Lodge, Costa Rica. (Photographed by Paul F.J. Eagles.)

dealing with cultural heritage, should increase in supply. Older people are much less likely to camp and much more likely to seek roofed accommodation such as lodges, hotels and cabins. Since most parks have a scarcity of such accommodation, this trend will reduce park use by older citizens or create a larger market for private accommodation providers near the park.

The baby boom generation is entering the retirement phase of life in large numbers early in the 21st century. This generation will be the healthiest, wealthiest and most numerous retirement population in history. While in their healthy, early senior years they can be expected to partake in large amounts of travel. All indications are that this travel will involve substantial amounts of nature-based travel. National parks and other forms of protected areas will be frequently selected as choice destinations. With appropriate levels of infrastructure, services and accommodation, parks have a very lucrative group of potential visitors. This group has the money to purchase park services, programmes and products. Managers could benefit from abundant levels of volunteer effort from many highly skilled people. The possibilities for donations of money are high. Conversely, without appropriate services, programmes and infrastructure, these seniors will spend their talents, money and time elsewhere. It is an important management decision whether to cater to the rapidly emerging market of seniors' tourism. If parks decide to not provide for these people's demands directly, they can expect to receive some of their use indirectly, through a third party. Private ecotourism ecolodges and tour operators are entering this seniors market aggressively and will bring many of their clients to parks and protected areas.

Trend 4. *As educational levels rise, demand for appreciative and learning opportunities associated with parks and protected areas will increase, resulting in implications for facilities, services and programmes*

Globally, there are higher levels of educational attainment. As people become more educated, their naturally inquisitive character is expressed in national parks and protected areas through desires for more information, interpretation and knowledge about the area and the values it contains. Park information must be adapted to a more sophisticated audience. This involves all aspects of information management, from websites to management plan contents, from resource policy to pricing policy. This also means that interpretive services must become more sophisticated in terms of what topics are discussed and how that information is delivered to an eager, willing and sophisticated audience. Since visitors are more knowledgeable in general and about parks in particular, interpreters will have to increase their skill and knowledge levels accordingly, thus requiring more formal education and training not only in the subject of the interpretive task but also in the technology and approaches to dealing with people. Many parks will experience higher levels of use by specialized ecotourism operators. These are private individuals providing programmes to a niche clientele. This will range from adventure travel experiences for youth to specialized nature education for retirees.

Trend 5. *Advances in the technology of travel and reductions in costs will result in increased demand for park and protected area opportunities distant from one's residence*

We have seen how changes in technology have affected international travel. International travel is expected to grow as dramatically in the early 21st century as it did in the late 20th century, thereby increasing the demand for national parks and protected areas distant from visitors' residences. The volume of air travel is expected to increase over the next 10 years as new aeroplane technologies come on-line, reducing the price of travel. By making their travel more affordable more people can visit foreign destinations. This trend means that park and protected area managers must be aware of travel trends for proper planning and must begin to consider how they communicate with people with different languages and cultural backgrounds. This communication not only includes language considerations but the differences in custom and tradition as well. Many managers will be faced with visitors who come from very different cultural backgrounds from those of the visitors in the past. This will bring many challenges in information provision, safety, health provision and supervision (Fig. 14.3).

After 2010, the emerging gap between light oil supply and demand will cause large price increases. The impacts of this are discussed in Trend 16.

Fig. 14.3. The management of high-risk recreation will continue as an important park management activity. The costs of search and rescue, the exposure to liability and negative environmental impacts will tend to restrict parks' ability to allow such activity. Jet boat, Hammer Gorge, New Zealand. (Photographed by Paul F.J. Eagles.)

Trend 6. *Continued growing sensitivity to environmental impacts of human activity leads to new and different roles of protected areas, ones away from traditional limited access sites, to multi-use sites similar to biosphere reserves*

The original North American ideal of a national park was a place that protected a pristine environment in which people were only visitors. It excluded permanent human habitation, except for park staff. While the national parks and protected areas designed with this ideal resulted in vast areas of protection, it is now clear that this form of preservation is not enough to guarantee adequate protection of all biodiversity, the human cultural heritage, enhancement of recreational opportunities or preservation of needed life support systems. At the same time, there is declining opportunity for designating relatively pristine environments devoid of people. This means that concerns about environmental impacts can be addressed not only through rules and regulations about pollution and land-use planning, but also with other forms of protected areas. These areas may be more similar to the concept of a biosphere reserve or the British National Park rather than the North American model of a national park. These new reserves may have widespread access, may contain human habitation and variable land uses, and may be used to learn more about how the human–environment interface works. These greenfield reserves contain a sophisticated combination of conservation and use. They may fill one approach to sustainable development.

Trend 7. *Growing knowledge about visitors, parks and their interactions leads to more sophisticated and effective methods of managing park-based tourism*

Our scientific understanding of human–environment interactions is increasing geometrically, particularly as we study processes that occur at larger spatial scales and longer temporal scales than we have in the past. Coupled with our ability to display and model these impacts with sophisticated geographical information systems, we are better able to address visitor-induced impacts, the effects of planned and unplanned management actions and new issues as they arise. Old models of planning may no longer be adequate with new scientific knowledge and increasing demands by the public for participation in planning. Planners will be more concerned with how scientifically complex ideas and methods are communicated to the public. They will be more concerned with integrating both public resources of knowledge and scientific sources of knowledge into planning processes.

Trend 8. *The global increase in park area, number of parks and park visitation leads to the outstripping of the capability of many park management institutions*

On one hand, the public increasingly demands more protection for areas worthy of national park or protected area status, and, thus, governments respond with new park creation. On the other hand, the public demands lower taxes. Governments respond with reduced public expenditures for park and protected area management. Increasing travel volumes result in increasing numbers of visitors to many parks and protected areas. The growing area to manage, the increasing level of visitation, and the decreasing budgets directly and negatively affect the institutional capacity of agencies to manage lands for which they have been mandated a stewardship responsibility. One result of a lack of adequate funding is the need for more personnel and personnel adequately trained to deal with increasing conflict in park management, new models of planning, and the new science and technology that is required to deal with increasing demand. In many park agencies these trends will lead to crisis levels of low managerial effectiveness. New approaches must be found to fund park management.

There is a real question, in addition, of whether the basic agency legislative mandates are appropriate for the new challenges, demand and conflicts of park management in the 21st century. For example, Wilkinson (1992) writes of the 'Lords of Yesteryear' referring to the laws of the late 19th century that apply to public lands in the western United States that are out of date but still in effect. These laws have severely hampered implementation of innovative natural resource planning processes.

Trend 9. *Park management shifts gradually from government agency structures, with centralized financial control, to parastatal forms, with financially flexible and increasingly democratic forms of management*
There is an increasing call for changed structures for many park management functions. This call is largely the result of the financial weaknesses of the typical government agency approach and changing political philosophies. In some places, this means contracting some park operations to private corporations, effectively replacing government employees and publicly-funded services. In other places, it means transferring some management functions to NGOs. In others, it means management by a park agency with a corporate organization, with management structures more similar to a corporation than a government agency. This parastatal form of management has a government-appointed committee to serve as the board of directors. The agency has wide abilities to earn, to retain income, to hire staff, to set prices and to operate. In essence the parastatal functions like a corporation owned by government. Some criticize this approach because of the possibility of motivation driven more by income generation than by public service. These management trends have tremendous and still unknown consequences for how cultural and natural values will be preserved, managed and presented to an increasingly demanding public. One of the most important areas of experimentation and research will be in the area of park administration and management.

Trend 10. *Park management funding increasingly shifts from government grants to park tourism fees and charges. This results in higher levels of visitor focus in management*

As government grants available for parks and protected areas do not keep up with the expansion of park area and park visitation, there is increased emphasis on the use of fees and charges on park visitors to provide the revenue necessary to fund park operations. This is widely known as the user pay approach (see Fig. 14.4).

Government policies in several countries provide a timetable for the park agency to replace some of, or the entire, park budget with earned income. For example, Parks Canada has a multiyear plan for increasing park income. However, the Government of Canada recognizes that it is not possible for national parks to earn all their financial needs from earned income. Therefore, the ultimate goal is to have a budget composed of income derived from both government allocations and earned income. In some countries with weak abilities of government to gain tax revenue, such as Tanzania and South Africa, the parks are not granted any government funding. All operational budgets must be income derived from fees and charges. We see the international trend as moving park management towards higher levels of earned income.

Parks have many potential sources of tourism-based income, including: entrance fees, recreation services fees, special events and special services, accommodation, equipment rental, food sales (restaurant and store), parking, merchandise sales (equipment, clothing, books, information, supplies), contractual agreements with concessionaires, licensing of intellectual property and cross-product marketing (Eagles, 2002). All of

Fig. 14.4. The relative amounts of park income coming from government grants and from park visitors will continue as a major policy debate in park management. Fee board, Zion National Park, USA. (Photographed by Paul F.J. Eagles.)

these sources are being used by some park agencies, but very few park agencies use the entire range. We expect that park agencies will utilize a much wider range of income sources in the future.

The trend for increased use of fee revenue has several implications for park management and the services delivered to visitors. One important implication is that of higher levels of charges for park services. Some people worry that this will keep some people from enjoying parks because of high cost. However, there is very little evidence of this trend in park use figures to date. Another implication is that the only benefits flowing from a park are those for which a charge exists. A third implication is that only services and opportunities that will break even between income and expenses will be provided.

If park operations are funded entirely by ongoing revenue from park visitors, the budget must stay in tune with projected revenues. If management costs increase there is the need to increase revenues. Sometimes the increased revenue may come from promotional campaigns designed to increase visitation, which in some cases may lead to adverse visitor impacts. In other cases, better pricing policy, the collection of fees from visitation that was formerly ignored, and higher fees associated with higher service levels may provide revenue that is sufficient to cover operating expenses. Increased fees can also raise expectations on the part of visitors about the quantity and quality of services that will be delivered as a result of the fee. Evidence from Canadian Provincial Parks shows that recent increases in fees are associated with higher use levels in parks. This counter-intuitive trend can be explained by the fact that the higher fees are associated with new, more efficient and better-targeted services. Therefore, park visitors are increasing their use of parks because they are better served by park management.

Parks with income derived from park visitation are much more client-oriented than parks utilizing government grants. Such parks are much more concerned about the visitor's length of stay, the visitor's satisfaction with the programmes and services, the visitor's recreational needs and the visitor's opinions about park management. This important financial trend in parks will spur considerable policy and management debate in parks in the future. For example, does the increased attention to the visitor come at the expense of attention towards biodiversity or cultural values? How much should a park be market driven or market responsive in deciding what services the park should offer?

Trend 11. *Parks and park agencies develop increased sophistication in their understanding and management of park visitation and tourism*

The implications for visitors of the growing availability of information, discussed earlier, also has parallel implications for park management. As databases about human populations become more widely accessible, park management has more information about potential visitors: the expectations the people bring with them, the lifestyles that different people live, the services they desire and their residence. This means that park and protected area managers can become much more sophisticated in tailoring management programmes and providing recreational opportunities to potential tourists. They may be able to deliver information ahead of the visit that will help to form appropriate expectations on the part of visitors. Park managers may be able to influence where within a park or among parks people visit. They may be able to design management programmes that can fine-tune visitor impacts and visitation patterns. Universities will be expected to increase their offerings in park tourism planning and management.

Trend 12. *Increasing globalization of information, business and government results in increasing international cooperation among park agencies, park visitors and non-governmental organizations regarding park management, especially in all aspects of park tourism*
The fact that there will be increased visitation to national parks and protected areas from all over the globe will be paralleled with increasing cooperation among park management agencies. This increased cooperation will occur in terms of trading information about park resources and opportunities. As in many sectors of business and government, park managers will travel more widely to learn from the experience of others. It will also involve information and management approaches dealing with visitors and tourists as well. Increased globalization and interconnections of industrial sectors may lead to increased cooperation between park management agencies and other sectors of the tourism industry such as airlines, hotels and tour-guide companies. In the future, park visitors and other concerned citizens will have more potential to work together on park management issues globally. Park managers can expect to deal with a more international clientele for all park programmes, including the planning processes that determine policy.

Trend 13. *Foreign aid and grants from NGOs increasingly fund biodiversity conservation and sustainable tourism development in developing nations. Typically, the goal is sustainable development that provides both conservation and economic benefit*
Where government funding does not keep up with park and tourism expansion, and the park institutional capacity to manage also decreases, it is likely that NGOs will expand their role not only in terms of funding and technical assistance but also in terms of direct management of parks and protected areas. For example, in the Netherlands Antilles, the Saba Conservation Foundation, and in Belize, the Belize Audubon Society, already undertake park management. In many parts of Africa, the ongoing efforts to develop sustainable tourism can be expected to pay off as increased ecotourism provides economic benefits to many parks.

As governments seek to find additional non-tax-based revenue to fund needed social programmes, NGOs may respond to potential resource commodity extraction programmes through an emphasis on sustainable development and management in protected reserves. As the expansion of protected areas in terms of their roles occurs, NGOs will provide assistance in developing programmes that more sensitively provide for the basic needs of people and biodiversity.

Trend 14. *Increasing demands for resource exploitation, such as oil, gas, minerals, water and wood, place stronger pressure for the exploitation of park resources in many locales. Healthy park tourism provides a counterbalance when tourism is sufficient to provide politically relevant benefits*
A growing world population and rising standard of living inevitably mean greater resource commodity extraction from a more limited and scarce water and land base. As governments attempt to find the resources to meet growing demands, they will look more and more frequently at national parks and protected areas as potential sources of exploitation. Some NGOs will advocate tourism as an alternative development tool in terms of providing jobs, labour income and tax revenues. NGOs will argue that park-based tourism is a better alternative than exploitation. Yet, some NGOs will also offer a sustainable development option, which may mean smaller-scale commodity extraction and processing that is focused on a land and water base that has higher levels of resilience.

Trend 15. *A substantial number of parks and their tourism will be destroyed by war, famine and civil unrest, especially in Africa and parts of Asia*
One of the facts of life which the human race has been unable to deal with successfully is conflict and war resulting from not only ethnic and religious strife but also the vast increases in human population numbers. Unfortunately some of this conflict will occur in or immediately adjacent to national parks and protected areas. Park management often ceases to exist in such times of conflict. For example, the Ugandan army wiped out the park service of Uganda during the Idi Amin years. This service never fully recovered from the loss of experienced people and equipment. The fall of the Shah of Iran spelled the end of Iran's National Park system. A change in government in Togo led to the complete loss of the national park system.

The Biwindi Impenetrable Forest National Park in Uganda was a site of terrorist activity against national park visitors in 2000. This action killed many people and destroyed a promising ecotourism industry. In addition, habitat for mountain gorillas in this part of Africa was partially destroyed as a result of the same regional conflict. There will be many more examples of individual parks being destroyed or damaged by local conflict.

If we accept the fact that some national parks will be impacted by war, famine and civil unrest, then perhaps park management agencies will develop contingency plans for the protection of any threatened or endangered species occurring in such places. There is some talk about the creation of an international conservation police force that has the mandate to enter and secure important national parks and protected areas during times of conflict. Such a force would be very controversial because of the perceived loss of sovereignty of national governments. As a result, it may take a long time to come into being.

Trend 16. *The world's international travel will continue to increase broadly until decreasing supplies of inexpensive light oil result in large increases in cost. As the energy costs increase, international travel will start to decline. Increasing stress on local resources will cause domestic economic stress resulting in substantial pressure and, most likely, damage to many parks and protected areas*

The world's prosperity in the 20th century was largely due to the abundant and inexpensive energy available from light oil. Inexpensive energy led to widespread travel. However, the Earth's supply of light oil is finite. As easily accessed oil fields become exhausted, more remote, deeper and harder to access supplies are being sought. The best estimates are that by 2010 the demand for light oil will exceed the supply (Campbell and Laherrere, 1998). From that time on, the difference between the supply and the demand will expand. This will have many implications. One is movement to increased use of heavy oil, coal, nuclear energy, natural gas and renewable energy. The other is much higher energy costs. When energy costs take off there will be dramatic changes in global consumption, economic and travel patterns. The implications for park tourism are considerable. Overall, international travel will decline in volume. Conversely, some domestic travel volume may increase, as people substitute local trips for longer voyages. Decreased economic vitality of many societies will result in severe pressures on many parks and protected areas. The recent experience in Russia may serve as a model. As the Russian economy declined in the latter years of the 20th century, local people accessing needed resources heavily exploited many Russian protected areas. Many sites rapidly transferred to other land uses, as civil stress enabled powerful groups to invade the protected areas for personal benefit. Very similar activities took place in Zimbabwe as the economy spiralled downwards and civil unrest rose.

As the world moves out of the era of abundant light oil, the impacts on park and protected area management in general and on park tourism specifically will be profound.

The increase in energy prices resulting from the divergence of light oil supply and demand could be the most significant trend affecting park tourism in the first 25 years of the 21st century.

Trend 17. *Global climate change will affect many parks and much park tourism*

Global climate change will be one of the most important environmental issues affecting parks and tourism in the 21st century. According to the

Intergovernmental Panel on Climate Change (2001), it is likely that there will be:

- higher maximum temperatures and more hot days over nearly all land areas;
- higher minimum temperatures, fewer cold days and frost days over nearly all land areas;
- reduced diurnal temperature range over most land areas;
- increased heat index over land areas;
- more intense precipitation events;
- increased summer continental drying and associated risk of drought in continental interiors;
- increased tropical cyclone peak wind intensities;
- increased tropical cyclone mean and peak precipitation intensities.

The implications are so far-reaching and significant that it is difficult to provide a succinct summary. However, a few trends are obvious. Globally the climate will change to a generally warmer climate. The impact will be stronger in the higher latitudes, with much warming in Arctic environments. Global climate change will reduce the ability of some parks to accept tourism, through drought, intense heat and rising ocean level. Other parks will have longer operating seasons, due to reduced seasonal impacts, such as winter closures. Increased tropical cyclone wind and precipitation intensity may cause severe damage to some parks, resulting in lowered attractiveness to visitation and lowered abilities to accept visitation. Regional impacts may be considerable. For example, increased drought and heat in the southern and central USA may cause heavy migration of people northwards, both permanently and seasonally. Northern US parks and Canadian parks may experience much higher visitation pressure as a result.

The implications of global climate change will be large and profound. All park planners and managers must consider these trends to their fullest extent. Some of the impacts can be dealt with under current management scenarios. Others will require entirely new approaches.

Roles of Park-based Tourism in the 21st Century

Given the above, tourism in national parks and protected areas will have expanded roles in the 21st century. To a great extent, these roles began their evolution in the late 20th century, but will grow and evolve over the next 25 years.

Role 1. *National parks and protected areas – including new classes of protected areas – become more integrated into regional and national economic development*

As we noted above, demands for national parks and protected areas coupled with a growing population, much of which seeks a higher quality of life, will stress the capacity of natural resources to produce needed goods and services. What this means for national parks and protected areas is that they will assume a wider variety of roles in landscape, cultural and biodiversity protection, but that these roles will tend to be integrated into needs for economic development (Fig. 14.5). As nations and regions within them build new economic development strategies, parks and protected areas that had formerly been excluded from such plans will be viewed as sources of jobs, income and tax revenue. This view will encompass not only the tourism potential of such parks and protected areas, but also the potential within them to carefully and sensitively provide resource commodities, in ways that do not replicate the large-scale landscape-changing methods of the past. The notions of sustainability and sustainable development as we briefly mentioned earlier in the book will drive the management plans for regional reserves and protected areas. For example, the Steens Mountains Cooperative Management and Protection Area in south-eastern Oregon includes objectives and landscapes to maintain the traditional culture and customs of a ranching industry. In this area, one of the criteria for management decisions is the viability of the ranching industry. Such sustainable development options in parks will be heavily debated in the coming years.

Fig. 14.5. The role of aboriginal peoples in park management will continue to be a major policy debate. The role of native peoples in the determination of park values, in the provision of park services and in park management may increase. Restored Indian village, Crawford Lake Conservation Area, Ontario, Canada. (Photographed by Paul F.J. Eagles.)

We would expect to see more actively managed biosphere reserves, composed of the traditional protected area in the core, and surrounding regions in which a variety of landscape management activities occur, with those activities focused on providing needed commodities for the local and regional human population. In the regions surrounding biosphere reserves, the guiding principles would be some kind of sustainable development strategy in which the local population has an important role in management decisions and where the timeframe for those decisions is relatively long. Newly established buffer areas near some Nepalese national parks may provide the prototype for such management.

Role 2. *National parks and protected areas assume larger roles in biodiversity protection, but those roles are directly related to new and creative ways of managing for natural resource commodities*
It is now quite clear that one of the most fundamental and persuasive rationales for designating an area as a national park or for other types of preservation is the potential biodiversity value that may be contained within it. To a very great extent, we do not completely understand the potential biodiversity and genetic resources contained within these areas but we do understand that, without some kind of protection, we may irretrievably lose these values yet be completely unaware that that has happened. At the same time, we also know that there are millions of people that expect to have a higher standard of living and thus in some way, in many protected areas some kind of resource commodity production will occur.

On the one hand, we know that we must preserve the biodiversity that we have inherited but on the other hand we also know that without curbs on population growth we will have to manage many of these places for some kind of production. What this means is that new protected areas will have to be more creatively managed to meet both goals, a job of such challenge and complexity that we probably cannot conceive of it now. It may very well be that the most fundamental economic value protected areas have in the 21st century is their value for the genetic and biological diversity they contain.

Role 3. *National parks and protected areas become model areas for sensitive forms of tourism development*
The interests in protecting biological diversity through designation of national parks and other types of protected areas converge with the need for economic benefits. In national parks, the second need can be accomplished through sensitive and appropriate forms of tourism development. The literature on sustainable and ecotourism development has grown dramatically since 1990. Yet, the authors have the distinct feeling that we do not have all the answers on how to create more sensitive tourism development.

National parks are a perfect place to demonstrate careful and sensitive tourism development, development that provides the recreation opportunities that visitors seek, the economic opportunities the local communities desire, and protection of biodiversity values that many people crave. National parks can assume a leadership role in demonstrating what sustainable tourism development is all about. This leadership role will itself require more creative leadership ideas within each national park agency, ideas that address the goals identified above but also recognize growing concerns about privatization of public resources.

Role 4. *Park-based tourism continues to emphasize learning as a major component, with innovative businesses assuming larger roles in this*

While many national parks and protected areas provide opportunities for outstanding recreation focusing on activities that present visitors with high levels of challenge and adventure, probably the greatest strength of national parks is their role in the environmental learning necessary to maintain the beautiful world that is our legacy. The combination of relatively pristine environments and modest levels of economic development activity – such as tourism – provide learning opportunities for all ages of the visiting public. By learning we mean much more than the kind of classical interpretive activities that national parks have engaged in in the past.

Learning in our sense is active, to some degree it is experimental, and it applies to visitors, the public living outside the park and the administrative organization. It is oriented towards not only understanding the natural processes which formed the landscape but also how we can better manage our activities and live within the constraints of these natural processes. We would expect to see private entrepreneurs developing opportunities to build businesses around this learning objective.

Role 5. *National parks and protected areas become a centre of learning about how people and natural processes interact*

A major dimension of the above role will be creating scientific opportunities and venues oriented around developing a better understanding of how people and natural processes interact. If there is one thing we have learned over the last half-century about the natural environment it is that the longer we ignore the presence of natural processes in different environments the fewer the options we have when confronted with problems. For example, the early fire exclusion policies developed in North America in fire-dependent ecosystems came back to haunt forest management agencies in the last 10 years. As fuels accumulated, fires grew in size and intensity, threatening the human communities that the fire exclusion policies were designed to protect. As part of this learning focus, parks will become a venue with facilities and accompanying programmes for people to better understand the scientific knowledge that has been generated. These venues have the opportunity of becoming safe, stimulating locations for the public to deliberate on complex problems of environmental management.

Role 6. *Parks continue to develop as cultural icons in local, national and international communities*
Of course, parks will continue in their traditional roles of providing opportunities for people to visit and better understand our cultural and natural heritage. As parks become more broadly known, many of them will become icons for various communities either at the national level or at the local level. To a great extent many parks may become associated with or become symbols of specific national identities. Many local communities that were initially antagonistic or suspicious of national park creation, develop higher levels of appreciation over time.

As parks become stronger international symbols there is stronger international pressure on management policies. This leads to more effective international programmes, with designations such as Ramsar Wetland, World Heritage and Biosphere Reserve becoming more widespread and accepted. Such international designations lead to higher levels of tourism as people recognize the sites as being globally significant. International travel and recognition lead to the concept that parks and protected areas are of universal significance, not just national significance. Therefore, national parks and national wildlife refuges will take on the stature of international parks and international wildlife refuges. Such a trend is a natural outgrowth of the global ecosystem concept of ecology and the global travel phenomenon. However, it will evoke substantial debate as the nation state becomes less important in the affairs of people.

Research, Planning and Management Needs

These trends and new roles will stretch the existing institutional capacity of all nations to adequately provide the necessary stewardship for their national parks and protected areas. This stewardship increasingly requires a solid foundation in the social and biophysical sciences; research that is not only looking at issues developing in the future, but is increasingly integrated among disciplines. We know that the longer we delay decisions about emerging social and environmental issues, the narrower the resulting decision space, the fewer the options, the less we will be able to meet competing needs for the resources and opportunities contained within national parks and protected areas.

While research provides the foundation on which decisions can be made, science only informs, it does not dictate. Local and experiential knowledge, too, are important legitimate sources of information. Planning systems must be increasingly developed to be more inclusive of those forms of knowledge and of the people affected by or interested in a park or protected area. These systems must be developed to integrate or accommodate (rather than balance) a variety of overlapping interests and values. This will be a large challenge, because planning has traditionally been conceived of as a technical process, conducted by privileged bureaucrats in environments that were opaque to viewers. New planning paradigms will emphasize openness, learning and consensus building, skills that can be translated into other arenas of citizen life.

Management, the activity of day-to-day decision making, will be challenged not only by the increased demands of a growing population, but also a stable capacity to meet these demands. For example, the camping capacity in the American national park system has hardly increased in the last 30 years, despite a 37% increase in the population of the USA. As the demand has increased relative to supply, more of the public, managers and scientists have increasingly questioned the carrying capacity of parks for recreation. While this concept, as an operational theory, has little validity, the general concern about the biophysical and social impacts of tourism remains an important and legitimate issue for managers. The challenge here is primarily in framing or defining the problem to be addressed in ways that lead to appropriate responses and management actions, in developing evaluation criteria that are more inclusive of the consequences of tourism management alternatives, and in being more creative about what management tools would be useful.

In summary, the practice of managing tourism in parks and protected areas has gone beyond the era of knee-jerk actions to become a sophisticated, inclusive, yet science-based profession. The values at stake in practising stewardship of these areas are often essential to the national identity of particular cultures. Management and science need to recognize this.

Conclusion

National park and protected area managers are confronted with numerous and seemingly increasingly intractable management problems. At the same time, they play incredibly important roles in our society in that they manage special places so our grandchildren will come to enjoy and appreciate them as we have. While managers might easily be overwhelmed by the seriousness and apparent complexity of park management problems, it is just as easy to see the tremendous and almost infinite opportunity that national parks and protected areas have for doing good in a world increasingly challenged by conflict and poverty.

Managers can more easily address the complex problems, and the tremendous opportunity for values they have been assigned to protect and the people for which they work, by implementing existing management systems such as those we have discussed in this text. Part of a protected area manager's job is to work with the people who benefit from these places so that they better understand the values contained within them and so that the manager better understands the needs of the people.

The trends that affect park and protected area management are difficult to predict precisely. This uncertainty can lead to anxiety and even stalemated action as people, including park managers, remain confused about the appropriate courses of action to initiate. While the future is difficult to predict, we can prepare for it. This preparation is founded on understanding management systems, the role of people and the principles of ecosystem process. It is through such a foundation, coupled with intelligent responsiveness, that managers can practise the stewardship with which they have been charged.

References

Campbell, C.J. and Laherrere, J.H. (1998) The end of cheap oil. *Scientific American* 278(3), 78–83.

Eagles, P.F.J. (2002) International trends in park tourism and economics. In: Bodrup-Nielsen, S. and Munro, N.W.P. (eds) *Managing Protected Areas in a Changing World: Proceedings of the Fourth International Conference on Science and Management of Protected Areas.* SAMPAA, Wolfville, Canada.

Intergovernmental Panel on Climate Change (2001) *Summary for Policymakers.* World Meteorological Organization, Geneva.

Wilkinson, C.F. (1992) *Crossing the Next Meridian: Land, Water, and the Future of the West.* Island Press, Washington, DC.

Further Reading

Gartner, W.C. and Lime, D.W. (2000) *Trends in Outdoor Recreation, Leisure and Tourism.* CAB International, Wallingford, UK.

Index

Page numbers in **bold** refer to material in tables, figures and boxes.